INTERNATIONAL DIRECTORY OF GEOSCIENCE ORGANIZATIONS

2nd Edition

Edited by

Nicholas H. Claudy
American Geological Institute
Education and Human Resources Department

U.S. Geological Survey
Reston, Virginia

American Geological Institute
Alexandria, Virginia

Cover

The GeoSphere® Image

This spectacular image of Earth marks a milestone in history. The image shows the Earth revealed, unobstructed by clouds. Through the use of NOAA weather satellites, hundreds of individual satellite images from space were combined to produce this image of the world. Drainage and relief features are enhanced. The composite, like a jigsaw puzzle, was assembled over a 10-month period and completed on April 15, 1990.

Satellite Composite View of Earth, by Tom Van Sant and The GeoSphere® Project, Santa Monica, California, with assistance from NOAA, NASA, Eyes on Earth. Technical direction is by Lloyd Van Warren. Source data derived from NOAA/TIROS-N Series Satellites. Completed April 15, 1990. All rights reserved by the GeoSphere® Project, 146 Entrada Drive, Santa Monica, CA 90402.

Staff

AGI:	Nicholas H. Claudy	Editor
	Dr. Xingbi Bai	Associate Editor (for China)
	Dr. Xiaobo Li	Associate Editor (for China)
	Lawrence Berg	Programmer
	Adrienne D. Fargas	Page Layout, Typesetting
	Veronika Litvak	Translation Services
	Liza Mallard	Page Layout, Typesetting
	Cathleen L. Piccione	Page Layout, Typesetting
	China Williams	*Geotimes*, Editorial Assistant
	Kay Yost	Page Layout, Typesetting

Printed by Capital City Press, Inc., Montpelier, Vermont.
Printed in the United States of America.

Published by the American Geological Institute.
Copyright © 1996 American Geological Institute; all rights reserved.

ISBN 0-922152-40-3

American Geological Institute
4220 King Street
Alexandria, VA 22302-1502
(703) 379-2480
FAX (703) 379-7563
ehrinfo@agi.umd.edu
http://www.agiweb.org/

TABLE OF CONTENTS

Preface v	ETHIOPIA 37
Introduction vi	FIJI 38
International Directory of Geoscience	FINLAND 38
Organizations	FRANCE 39
AFGHANISTAN 1	FRENCH GUIANA 41
ALBANIA 1	GABON 41
ALGERIA 1	GAMBIA 41
ANGOLA 2	GERMANY 41
ARGENTINA 2	GHANA 42
ARMENIA 3	GREECE 43
ARUBA 3	GREENLAND 43
AUSTRALIA 3	GUADELOUPE 43
AUSTRIA 9	GUATEMALA 43
BAHAMAS 10	GUINEA 44
BANGLADESH 10	GUINEA-BISSAU 44
BARBADOS 10	GUYANA 44
BELGIUM 11	HAITI 44
BELIZE 11	HONDURAS 45
BENIN 12	HONG KONG 45
BHUTAN 12	HUNGARY 46
BOLIVIA 12	ICELAND 47
BOTSWANA 13	INDIA 48
BRAZIL 13	INDONESIA 49
BRUNEI DARUSSALAM 14	IRAN 50
BULGARIA 15	IRAQ 51
BURKINA FASO 16	IRELAND 51
BURMA (see MYANMAR) 16	ISRAEL 52
BURUNDI 16	ITALY 53
CAMEROON 16	JAMAICA 54
CANADA 17	JAPAN 55
CHAD 22	JORDAN 60
CHILE 23	KAZAKHSTAN 60
CHINA 23	KENYA 60
COLOMBIA 31	KIRIBATI 61
CONGO 31	KOREA, DEMOCRATIC PEOPLE'S RE-
COOK ISLANDS 31	PUBLIC OF 61
COSTA RICA 32	KOREA, REPUBLIC OF 61
COTE D'IVOIRE 32	KUWAIT 62
CUBA 33	LAO PEOPLE'S DEMOCRATIC REPUB-
CYPRUS 33	LIC 62
CZECH REPUBLIC 34	LEBANON 62
DENMARK 34	LESOTHO 62
DJIBOUTI 34	LIBERIA 63
DOMINICA 35	LIBYAN ARAB JAMAHIRIYA 63
DOMINICAN REPUBLIC 35	LIECHTENSTEIN 63
ECUADOR 35	LITHUANIA 64
EGYPT 36	LUXEMBOURG 64
EL SALVADOR 37	MACEDONIA, REPUBLIC OF 64
ESTONIA 37	MADAGASCAR 64

TABLE OF CONTENTS

MALAWI	65
MALAYSIA	65
MALI	66
MALTA	66
MARSHALL ISLANDS	66
MARTINIQUE	66
MAURITANIA	66
MAURITIUS	67
MEXICO	67
MICRONESIA, FEDERATED STATES OF	69
MONGOLIA	69
MOROCCO	69
MOZAMBIQUE	69
MYANMAR	70
NAMIBIA	70
NEPAL	71
NETHERLANDS	71
NETHERLANDS ANTILLES	72
NEW CALEDONIA	72
NEW ZEALAND	72
NICARAGUA	73
NIGER	74
NIGERIA	74
NORWAY	75
OMAN	75
PAKISTAN	76
PANAMA	76
PAPUA NEW GUINEA	77
PARAGUAY	77
PERU	77
PHILIPPINES	79
POLAND	79
PORTUGAL	80
QATAR	81
REUNION	81
ROMANIA	82
RUSSIAN FEDERATION	82
RWANDA	86
SAMOA	87
SAUDI ARABIA	87
SENEGAL	87
SIERRA LEONE	88
SINGAPORE	88
SLOVAKIA	89
SOLOMON ISLANDS	89
SOMALIA	89
SOUTH AFRICA	90
SPAIN	91
SRI LANKA	93
SUDAN	93
SURINAME	94
SWAZILAND	94
SWEDEN	94
SWITZERLAND	95
SYRIAN ARAB REPUBLIC	96
TAIWAN, REPUBLIC OF CHINA	96
TANZANIA, UNITED REPUBLIC OF	98
THAILAND	98
TOGO	100
TONGA	100
TRINIDAD AND TOBAGO	100
TUNISIA	101
TURKEY	101
UGANDA	102
UKRAINE	102
UNITED ARAB EMIRATES	103
UNITED KINGDOM	103
UNITED STATES	108
URUGUAY	130
UZBEKISTAN	130
VANUATU	130
VENEZUELA	130
VIET NAM	131
YEMEN	132
YUGOSLAVIA	132
ZAIRE	133
ZAMBIA	133
ZIMBABWE	133
Organization Index	**135**

PREFACE

More than twenty years have passed since the U.S. Geological Survey (USGS) published the original *Worldwide Directory of National Earth-Science Agencies* as Geological Survey Circular 716. Although the original directory was intended primarily for internal USGS use, the demand for it both within and outside the USGS was far greater than expected, and the original printing was quickly consumed. To accommodate the continuing demand for a ready reference on international geoscience organizations, the USGS and the American Geological Institute (AGI) compiled the *International Directory of Geoscience Organizations*; the initial 1994 printing of this publication has now also been exhausted. In releasing this updated version of the *Directory*, the USGS and AGI have endeavored to incorporate the many changes that have occurred in the global geoscience community since the preparation of the original edition.

Paul P. Hearn, Jr.
Chief, International Programs
Geologic Division
U.S. Geological Survey

INTRODUCTION

The *International Directory of Geoscience Organizations, 2nd Edition*, presents a combination of three separate compilations that represent a total of 1,765 entries in 168 countries.

- The first and major source, previously published as *Worldwide Directory of National Earth-Science Agencies*, provides information concerning governmental geoscience agencies around the world which have functions similar to those of the U.S. Geological Survey (USGS);

- The second source is the *Directory of Geoscience Organizations*, which appears in the annual October issue of *Geotimes*, published by the American Geological Institute (AGI). This source adds international professional geoscience societies and organizations.

- The third source is the listing of foreign geoscience departments which formerly appeared in AGI's *Directory of Geoscience Departments*.

Data presented here are current as of August 1996. Much of the information about international governmental geoscience agencies was supplied by staff members of U.S. Embassies worldwide. Additional data were also provided by members of international agencies and professional geoscience societies. We are indebted to all those officials, to all USGS colleagues, and to others who have assisted us in providing information.

The principal functions of the governmental geoscience agencies are indicated by code letters, which appear in parentheses after an organization name. The Principal Function Index lists organizations in these categories.

(C) = Cartography

(E) = Environment

(G) = Geology

(H) = Hydrology

(R) = Resources Regulation

The *International Directory of Geoscience Organizations* is arranged alphabetically by country name and alphabetically within country. When provided by an agency or society, telephone, telex, telefax numbers, e-mail addresses, and World Wide Web addresses have been indicated as well as country and city codes (in parentheses). An index by organization name follow the main listings.

The country names used in the *International Directory of Geoscience Organizations* are in accordance with those set by the International Organization for Standardization (ISO), which is a worldwide federation of national standards bodies.

AFGHANISTAN

Afghan National Petroleum Company (G)
Ministry of Mines and Industries
Darulaman
Kabul

Department of Mines and Geology (G,H)
Ministry of Mines and Industries
Darulaman
Kabul
Minister: Noor Ahmad Azimi
Phone: 25848

ALBANIA

Albanian Geological Survey (G)
Ministry of Mining, Mineral and Energy Resources
Tirana
General Director: Vasil Grillo
Phone: (355-42)25580
Fax: (355-42)34052/34031
Telex: 604 4204 MIMEN AB

Association of Scientific Workers of the People's Republic of Albania (C)
Tirane
President: Kol Paparisto

Instituti Hidrometeorologjik (H)
Rruga "Durresi" 219
Tirane
Phone: (355-42)23518
Fax: (355-42)23518

ALGERIA

Centre National de Recherche et d'Applications des Geosciences (G)
Universite d'Alger
2, rue Didouche Mourad
Alger
Directeur: Nacer-Eddine Kazi-Tani

Centre National des Zones Arides (H)
Beni Abbes - Bechar
Directeur: Mme N. Bounaga

Direction des Mines et de la Geologie (G)
Ministere de l'Industrie Lourde
1 Rue Ahmed Bey de Constantine
Immeuble "le Colysee"
Algiers
Directeur: M. Mohamed Ramdani
Fax: (213)591243
Telex: 66190 MILD DZ

Ecole Nationale de Sciences Geodesiques (C)
B.P. 13
Arzew
Directeur: Rahel Redjouane

Institut d'Hydraulique (H)
Universite des Sciences et de la Technologie d'Oran
B.P. 1505
El-M'Nouar - Oran

Institut d'Hydraulique (H)
Universite des Sciences et de la Technologie Houari Boumediene
B.P. 9
Dar El Beida - Alger

Institut d'Hydrotechnique et de Bonification (H)
B.P. 31
Blida

Institut de Biologie et des Sciences de la Terre (G,H)
Universite d'Oran
Es-Senia - Oran

Institut des GeoSciences (G)
Universite de Sciences et de la Technologie d'Oran
B.P. 1505
El-M'Nouar - Oran

Institut des Sciences de la Terre (G,H)
Universite de Constantine
Ain El Bey - Constantine

Institut des Sciences de la Terre (C,G)
Universite des Sciences et de la Technologie Houari Boumediene
B.P. 9
Dar Al Beida - Alger
Directeur: Mohamed Tefiani

Institut National de Cartographie (C)
123 rue de Tripoli
B.P. 32 Hussein-Dey
Algiers
Director: Halim Mansour

Ministere de l'Hydraulique (H)
Ex-Grand Seminaire
Kouba, Algiers

1

Minister: Brahim Brahimi
National Office of Geology
Office National de la Geologie
18A Avenue Mustapha El Ouali
Algiers
Phone: (213) 61-42-70

Secretariat d'Etat aux Forets et a la Mise en Valeur des Terres (H)
Immeuble des Forets
Ex. Bois de Boulogne
El Madania, Alger
Secretaire d'Etat: Mohamed Rouighi

ANGOLA

Direccao de Servicos de Geologia de Angola (C,G)
Caixa Postal 1260-C
Luanda
Director: Dr. A. G. de Araujo

Observatorio Meteorogico e Magnetico (G)
Joao Capelo
Rua Diogo Cao 20
Caixa Postal l288-C
Director: Alberto Leao Diniz
Telex: 317 0ENE AN

ARGENTINA

Argentine Association of Mineralogy, Petrology and Sedimentology
Pabellon 2, Piso 1
Ciudad Universitaria-Nunex
1428 Buenos Aires

Argentine Geological Association
Asociacion Geologica Argentina
Maipu 645
Piso 1
1006 Buenos Aires
A.C. Riccardi
Phone: (54-1) 325-3104
Fax: (54-1) 325-3104
E-Mail: aga@aga.uba.edu.ar

Argentine Geological Survey
Secretaria de Mineria de la Nacion
Avenida Julio A. Roca 651
1322 Capital Federal
Buenos Aires
Phone: (54-1) 343-3683
Fax: (54-1) 343-3525

Argentine Paleontological Association
Asociacion Paleontologica Argentina
Maipu 645
Piso 1
1006 Buenos Aires
E.J. Romero
Phone: (54-1) 322-2820

Commission on the Quaternary of South America
International Union for Quaternary Research
cc 92
9410 Ushuaia
Tierra del Fuego
Jorge Rabassa
Phone: (54) 901-22310/12/14
Fax: (54) 901-22318

Direccion de Inversiones Mineras
Julio A. Roca 651
9 piso
1322 Capital Federal
Buenos Aires
David Borelli
Phone: (54-1) 349-3232
Fax: (54-1) 349-3236
E-Mail: dborel@secind.mecon.ar

Gerente División Exploración (G)
Yacimientos Petroliferos Fiscales
Avenida Rogue Sáenz Pena 777
1364 Buenos Aires
Engineer: Mateo Alberto Turic
Phone: (54-1)329-2136
Fax: (54-1)329-5643
E-Mail: 73070.1044@compuserve.com

Instituto de Geocronologia y Geologia Isotopica (G,H)
Pabellón INGEIS
Ciudad Universitaria
1428 Buenos Aires
Director: Enrique Linares
Phone: (54-1) 783 3021
Fax: (54-1) 783 3024
E-Mail: postmaster@ingeis.uba.ar

Instituto Geografico Militar (C)
Avenida Cabildo 381
1426 Buenos Aires
Director: Horacio Esteban Avila
Phone: (54-1)7713031
Fax: (54-1)7761611

Instituto Nacional de Ciencia y Tecnica Hidricas (H)
767 Lima Piso 1
1037 Buenos Aires

Presidente: Alberto J. Calamante
Phone: (54-1)3340891/0724

Ministerio de Economia y Obras y Servicios Publicos (G)

Secretaria de Mineria
Buenos Aires
Secretary of Mining: Dr. Angel Eduardo Maza
Fax: (54-1)331226
National Director of Mining: Dr. Miguel Angel Guerrero

National Director of the Metal Mining Economy

Director Nacional Economia Minera
Avenida Santa Fe 1548
Piso 5
1060 Buenos Aires

National Mining Service

Servicio Minero Nacional
Avenida Santa Fe 1548
Piso 4
1060 Buenos Aires

Secretaria de Estado de Ciencia y Tecnologia (G)

Cordoba 831, 2 piso
1054 Buenos Aires
Secretary: Dr. Manuel Sadosky
Phone: (54-1)3121706/8364
Telex: 25272 SECYT AR

Servicio de Hidrografia Naval (H)

Montes de Oca 2124
1271 Buenos Aires
Chief: Capitan de Navio Manuel Guillermo Videla
Phone: (54-1)216884/0061/7797/4175

Servicio Geológico Nacional (C,E,G)

Dirreción de Geologia Regional
Secretaria de Mineria e Industria
Avenida Julio A. Roco 651 - Piso 10
1322 Buenos Aires
Director: Jose E. Mendia
Phone: (54-1)349-3186
Fax: (54-1)349-3119

Yacimientos Carboniferos Fiscales (G)

Avenida Roque Saenz Pena 1190,
20 piso
1035 Buenos Aires
Matar Ibanez
President: Ing. Jose Agustin
Phone: (54-1)34001/7211
Telex: 18126 YACAFI AR

ARMENIA

Institute of Geology (G)

24A Bagramian Ave.
Yerovan 375019
Director: Sergei Y. Grigorian
Phone: 524426

ARUBA

Department of Geodesy and Realty Registration (DLV) (G)

J.E. Irausquinplein 10
Oranjestad
Director: Ing. Vivienno L. Frank
Phone: (297-8)21311
Fax: (297-8)36905

Department of Public Works (DOW) (E,G,H)

Paardenbaaistraat 2
Oranjestad
Director: Hose B.Th. Figaroa
Phone: (297-8)24700
Fax: (297-8)38003

AUSTRALIA

University of Adelaide

Department of Geology & Geophysics
GPO Box 498
Adelaide
South Australia 5005
Phone: 08-303-5375/6
Fax: 08-303-4347

National Key Centre for Petroleum Geology & Geophysics
Thebarton Campus
Adelaide
South Australia 5505
Phone: 08-303-4299
Fax: 08-303-4345
E-Mail: wstuart@pet2.ncpgg.adelaide.edu.au

Agriculture and Resource Management Council of Australia and New Zealand

Rural Division
Dept of Primary Industries and Energy
GPO Box 858
Canberra, ACT 2601
John Graham

Phone: (61-62) 272 5093
Fax: (61-62) 272 4772

Association of Australasian Paleontologists
Geological Society of Australia
1203 Wynyard House
301 George St.
Sydney, New South Wales 2000
Phone: (61-2) 290 2194
Fax: (61-2) 290 2198
E-Mail: misha@gsa.org.au

Australasian Institute of Mining and Metallurgy
P.O. Box 660
Carleton South
Victoria 3053
J.M. Webber
Phone: (61-3) 9662 3166
Fax: (61-3) 9662 3662
http://www.ausimm.com.au

Australian Geological Survey Organisation
GPO Box 378
Canberra
Australian Capital Territory 2600
Neil Williams
Phone: (61-6) 249-9111
Fax: (61-6) 249-9999
http://www.agso.gov.au/

Australian Geomechanics Society
Institution of Engineers
11 National Circuit
Barton
Australian Capital Territory 2600
Peter May
Phone: (61-62) 270-6555
Fax: (61-62) 273-1488

Australian Geoscience Information Association Inc.
Library Services
Esso Australia Ltd.
GPO Box 400C
Melbourne Victoria 3001
Julie Adams
Phone: (61-3) 270 3829
Fax: (61-3) 270 3994

Australian Institute of Geoscientists
22 Kurraba Road
Neutral Bay
New South Wales 2089
Lindsay N. Ingall
Phone: (61-2)9955 1788
Fax: (61-2)9957 6788

Australian Mineral Foundation Inc.
63 Conyngham St.
Glenside
South Australia 5065
Phone: (61-8) 379 0444
Fax: (61-8) 379 4634
http://www.amf.com.au/amf

Australian National University
Centre for Australian Regolith Studies
G.P.O. Box 4
Canberra
Australian Capital Territory 0200
Phone: 06-249-3406
Fax: 06-249-0738

Department of Geology
Canberra
Australian Capital Territory 0200
Phone: 06-249-2056
Fax: 06-249-5544
E-Mail: mjr653@cscgpo.anu.edu.au

Australian Society of Exploration Geophysicists
411 Tooronga Road
Hawthorn East Victoria 3123
Janinie Cross
Phone: (61-3) 9822-1399
Fax: (61-3) 9822-1711
E-Mail: aseg@unimelb.edu.au

Ballarat University College
Department of Geology
P.O. Box 663
Ballarat
Victoria 3353
Phone: 053-279-260
Fax: 053-339-105

Remote Sensing Unit
P.O. Box 663
Ballarat
Victoria 3353
Phone: 053-339-260
Fax: 053-339-105

University of Canberra
Department of Geology
P.O. Box 1
Belconnen
Australian Capital Territory 2616
Phone: 06-252-2525
Fax: 06-252-2166

Centre for Ore Deposit and Exploration Studies (CODES)
University of Tasmania
GPO Box 252C
Hobart 7001
Ross Large

Phone: (61) 0220 2472
Fax: (61) 0220 7662
E-Mail: ross.large@geol.utas.edu.au

Commonwealth Scientific and Industrial Research Organization (CSIRO) (G,H)

Institute of Minerals, Energy and Construction
North Ryde Laboratories
P.O. Box 93
North Ryde 2113
Director: A.F. Reid
Phone: (61-2)8878241
Telex: AA25817

Curtin University of Technology

Australian Petroleum Co-operative Research Centre
G.P.O Box U1987
Perth
Western Australia 6001
Phone: 09-351-7510
Fax: 09-351-2377

Co-operative Research Centre for Australian Mineral Exploration Technologies
G.P.O. Box U1987
Perth
Western Australia 6001
Phone: 09-351-7510
Fax: 09-351-2377

Department of Exploration Geophysics
GPO Box U 1987
Perth
Western Australia 6001
Phone: 09-351-7510
Fax: 09-351-2377
E-Mail: nthompson@cc.curtin.edu.au

Department of Mineral Exploration & Mining Geology
WA School of Mines
Curtin University, Kalgoorlie
P.O. Box 597
Kalgoorlie
Western Australia 6430
Phone: 090-220-132
Fax: 090-911-400

National Key Centre for Resource Exploration
G.P.O. Box U1987
Perth
Western Australia 6001
Phone: 09-351-7968
Fax: 09-351-3153

School of Applied Geology
G.P.O. Box U1987
Perth
Western Australia 6001
Phone: 09-351-7968
Fax: 09-351-3153

WA Centre for Petroleum Exploration
G.P.O. Box U1987
Perth
Western Australia 6001
Phone: 09-351-7968
Fax: 09-351-3153

Department of Mineral Resources (G,R)

Minerals and Energy House
29-57 Christie St.
Box 536
St. Leonards
New South Wales 2065
Phone: (61-2)9901-8888
Fax: (61-2)9901-8777

Department of Minerals and Energy

100 Plain St.
East Perth
Western Australia 6004
Phone: (61-9) 222-3333
Fax: (61-9) 222-3510

Department of Mines and Energy South Australia

Box 151
Eastwood
South Australia 5063
A. Andrejewskis
Phone: (61-8) 274 7500
Fax: (61-8) 272-7597
E-Mail: andrejewskis, andrew@msgate.mesa.sa.gov.au

Dept of Primary Industries and Energy (C,G,H)

Australian Geological Survey Organization (AGSO)
GPO Box 378
Canberra
Australian Capital Territory 2601
Minister DPIE: John Anderson
Phone: (61-6)2499111
Fax: (61-6)2499999
http://www.agso.gov.au

Bureau of Resource Sciences (BRS)
John Curtin House
22 Brisbane Avenue
Bartony, Australian Capital Territory 2600
Director: Harvey Jacka
Phone: (61-6)249-9111
Fax: (61-7)574614
Telex: AA62109
Executive Director: Neill Williams
Phone: (61-7)499600

Economic Geology Research Unit
James Cook University of North Queensland
Townsville Q4811
Dee-Ann Bely
Phone: (61-77) 814726
Fax: (61-77) 251501
E-Mail: dee-ann.cany@jcu.edu.au

Federation of Australian Scientific and Technological Societies
GPO Box 2181
Canberra
Australian Capital Territory 2601
David Widdup
Phone: (61-77) 247 3554
Fax: (61-77) 249 6419

Flinders University of SA
School of Earth Sciences
G.P.O. Box 2100
Adelaide
South Australia 5001
Phone: 08-201-2212
Fax: 08-201-2676
E-Mail: esch@flinders.edu.au

Geological Society of Australia
1203 Wynyard House
301 George St.
Sydney
New South Wales 2000
Phone: (61-2) 290 2194
Fax: (61-2) 290 2198
E-Mail: misha@gsa.org.au

Geological Survey Division
Department of Mines and Energy
GPO Box 194
Brisbane
Queensland 4001
G.W. Hofmann
Phone: (61-7) 3237 1499
Fax: (61-7) 3235 4074
E-Mail: dme@mailbox.uq.oz.au

Geological Survey of Victoria
Department of Energy and Minerals
P.O. Box 2145 MDC
Fitzroy
Victoria 3065
T.W. Dickson
Phone: (61-3)9412 7802
Fax: (61-3)9412 7803

Geological Survey of Western Australia
Department of Minerals and Energy
100 Plain St.
East Perth
Western Australia 6004
P. Guj
Phone: (61-9) 222-3333
Fax: (61-9) 222-3633

Institute of Minerals, Energy and Construction
Box 93
North Ryde
New South Wales 2113
Phone: (61-2) 887-8200
Fax: (61-2) 887-8197

International Association of Volcanology and the Chemistry of the Earth's Interior (IAVCEI)
Australian Geological Survey Organization
GPO Box 378
Canberra, Australian Capital Territory 2601
R.W. Johnson
Phone: (61-6) 249-9377
Fax: (61-6) 249-9986
http://www.-geo.lanl.gov/heiken/iavcei_home_page

James Cook University of North Queensland
Coalseam Gas Research Institute
Townsville
Queensland 4811
Phone: 077-81-5195
Fax: 077-251-501
E-Mail: gl.pjc@jcu.edu.au

Department of Earth Sciences
Townsville
Queensland 4811
Phone: 077-814-546
Fax: 077-251-501
E-Mail: geology@jcu.edu.au

Economic Geology Research Unit
Townsville
Queensland 4811
Phone: 077-814-726
Fax: 077-251-501

National Key Centre in Economic Geology
Townsville
Queensland 4811
Phone: 077-81-4726
Fax: 077-251-501

Key Centre for Teaching and Research in Strategic Mineral Deposits
Department of Geology and Geophysics
University of Western Australia
Nedlands
Western Australia 6907
David I. Groves

Australia
International Directory of Geoscience Organizations
Australia

Phone: (61-9) 380-2667
Fax: (61-9) 380-1178
E-Mail: jthicket@geol.uwa.edu.au

Latrobe University College of Advanced Education
Department of Geology
P.O. Box 199
Bendigo
Victoria 3550
Phone: 054-447-222
Fax: 054-447-777

Latrobe University
School of Earth Sciences
Bundoora
Victoria 3083
Phone: 03-479-2649
Fax: 03-479-1272

Macquarie University
School of Earth Sciences
North Ryde
New South Wales 2109
Phone: 02-850-8418
Fax: 02-850-8428

University of Melbourne
School of Geology
Parkville
Victoria 3052
Phone: 03-344-6520
Fax: 03-344-7761

Mineral Resources Tasmania
Box 56 Rosny Park Tas
Tasmania 7018
Phone: (61-0362) 338-333
Fax: (61-0362) 338-338
E-Mail: tbrown@tdr.tas.gov.au

Monash University
Department of Earth Sciences
Clayton
Victoria 3168
Phone: 03-565-4879
Fax: 03-565-4903

University of New England
Department of Geology & Geophysics
Armidale
New South Wales 2351
Phone: 067-732-860
Fax: 067-712-898
E-Mail: bmckelve@metz.une.edu.au

University of New South Wales
Department of Applied Geology
P.O.Box 1
Kensington
New South Wales 2033
Phone: 02-697-4285
Fax: 02-313-8883

Groundwater Management & Hydrogeology
P.O. Box 1
Kensington
New South Wales 2033
Phone: 02-697-4275
Fax: 02-662-1923

New South Wales Key Centre for Mines
P.O. Box 1
Kensington
New South Wales 2033
Phone: 02-697-5006
Fax: 02-313-7269

University of Newcastle
Department of Geology
Callaghan
New South Wales 2308
Phone: 049-215-411
Fax: 049-216-925
E-Mail: glgam@cc.newcastle.edu.au

Institute of Coal Research
Callaghan
New South Wales 2308
Phone: 049-215-400
Fax: 049-215-925

Northern Territory Geological Survey
Box 2901
Darwin
Northern Territory 0801
Paul Le Messurier
Phone: (61-89) 89 5511
Fax: (61-89) 89 6824

University of Queensland
Department of Earth Sciences
Brisbane
Queensland 4067
Phone: (61-7) 3652375
Fax: (61-7) 3651277
E-Mail: earthsci@uqvax.cc.uq.edu.au

Queensland University of Technology
School of Geology
G.P.O. Box 2434
Brisbane
Queensland 4001
Phone: 07-864-2324
Fax: 07-864-1535

Royal Melbourne Institute of Technology
Department of Civil & Geological Engineering
G.P.O. Box 2476V
Melbourne
Victoria 3001

Phone: 03-660-2208
Fax: 03-639-0138

Phillip Institute of Technology, Seismology Research Centre
Plenty Road
Bundoora
Victoria 3083
Phone: 03-468-2468
Fax: 03-467-6184

University of South Australia

Department of Applied Geology
Gartrell School of Mining, Metallurgy & Applied Geology
Ingle Farm
South Australia 5095
Phone: 08-302-3107
Fax: 08-302-3378

Department of Geology
Smith Road
Salisbury East
South Australia 5109
Phone: 08-302-5137
Fax: 08-302-5101

South Australian Department of Mines and Energy

Geological Survey of South Australia
Box 151
Eastwood
South Australia 5063
Chief Executive Officer: Andrew Andrejewskis
Phone: (61-8) 274-7500
Fax: (61-8) 272-7597
E-Mail: aandrejs@msgate.mesa.sa.gov.au

Southern Cross University

Faculty of Resource Science & Management
P.O. Box 157
Lismore
New South Wales 2480
Phone: 066-203-650
Fax: 066-212-669

Sustainable Land and Water Resources Management Committee

Department of Primary Industry and Energy
Land Resources Division
GPO 858
Canberra, Australian Capital Territory 2601
Secretary: R. Waldron
Phone: (61-6)2725028
Fax: (61-6)2724526
Telex: AA62188

University of Sydney

Department of Geology & Geophysics
Sydney
New South Wales 2006
Phone: 02-692-3244
Fax: 02-692-0184

Ocean Sciences Institute
Sydney
New South Wales 2006
Phone: 02-692-3244
Fax: 02-692-0194

University of Tasmania

Antarctic CRC
G.P.O. Box 252C
Hobart
Tasmania 7001
Phone: 002-207-888
Fax: 002-202-973
E-Mail: l.nielsen@antrc.utas.edu.au

Department of Geology
G.P.O. Box 252C
Hobart
Tasmania 7001
Phone: 002-202-476
Fax: 002-232-547
E-Mail: j.hankin@geol.utas.edu.au

National Key Centre for Ore Deposits & Exploration Studies
G.P.O. Box 252C
Hobart
Tasmania 7001
Phone: 002-202-476
Fax: 002-207-662
E-Mail: ross.large@geol.utas.edu.au

University of Technology, Sydney

Department of Applied Geology
P.O. Box 123
Broadway
New South Wales 2007
Phone: 02-330-1763
Fax: 02-330-1755

University of Western Australia

Department of Geology
Nedlands
Western Australia 6009
Phone: (09) 380-2666
Fax: (09) 380-1037
Telex: AA92992
E-Mail: caroline@geofiz.geol.uwa.edu.au

National Key Centre for Strategic Mineral Deposits
Nedlands
Western Australia 6000
Phone: 09-380-2667
Fax: 09-380-1037

University of Wollongong
Department of Geology
PO Box 1144
Wollongong
New South Wales 2522
Phone: 042-213-841
Fax: 042-214-250
E-Mail: b.mcgoldrick@uow.edu.au

AUSTRIA

Austrian Geological Society
Osterreichische Geologische Gesellschaft
Rasumofskygasse 23
Box 127
A-1031 Vienna
Secretary General: Johann Egger, Thomas Hofmann
Phone: (43-1) 712 56 74-0
Fax: (43-1) 712 56 74 56
E-Mail: jeeger@cc.geolba.ac.at

Austrian Mineralogical Society
Osterreichische Mineralogische Gesellschaft
Naturhistorisches Museum
Postfach 417
A-1014 Vienna
Franz Brandstatter
Phone: (43-1) 52 177, EXT. 270
Fax: (43-1) 52 177 264

Austrian Paleontological Society
Institut fur Palaontologie de Universitat
Geozentrum
Althanstr. 14
A-1090 Vienna

Central Institute for Meteorology and Geodynamics (G,H)
Hohe Warte 38
P.O.B. 342
A-1190 Vienna
Director: Peter Steinhauser
Phone: (43-1)36 0 26
Fax: (43-1)36 91233
Telex: 131837 A METWA
http://www.univie.ac.at/zamg/

Federal Authority for Standardization and Geodesy (C)
Gruppe Landesaufnahme
Schiffamtsgasse 1-3
A-1025 Vienna
President: Friedrich Hrbek
Phone: (43-1))357611
Fax: (43-1)2161062
Telex: 114568

Federal Environment Agency (E,H)
Spittelauer Lande 5
A-1090 Vienna
Director: Wolfgang Struwe
Phone: (43-1)31304-0
Fax: (43-1)31304-5400
http://www.ubavie.gr.at

Federal Institute for Testing and Research/Geotechnical Institute (G,H)
Faradaygasse 3
P.O.B. 8
A-1031 Vienna
Director: Diets Sauer
Phone: (43-1)79747-496
Fax: (43-1)79747-592
Telex: 136677
http://www.arsenal.ac.at

Federal Office and Research Institute of Agriculture (C,E,G)
Spargelfeldstrasse 191
A-1226 Vienna
Director: Otto Danneberg
Phone: (43-1)28816-2000
Fax: (43-1)28816-2106

Geological Survey of Austria (G)
Rasumofskygasse 23
P. O. B 127
A-1031 Vienna
Acting Director: W.R. Janoschey
Phone: (43-1)71256740
Fax: (43-1)712567456
Telex: 132927

International Society of Soil Science
Universitaet fuer Bodenkultur
Gregor Mendel-Strasse 33
A-1180 Vienna
Winfried E.H. Blum
Phone: (43-1) 310-6026
Fax: (43-1) 310-6027
E-Mail: isss@edv1.boku.ac.at

Ministry of Agriculture and Forestry/Hydrographical Central Bureau (H)
Marxergasse 2
A-1030 Vienna
Head: Franz Nobilis
Phone: (43-1))71100
Fax: (43-1)711006851

BENIN

Centre National de la Recherche Appliquee (G)
Ministere de L'Enseignement de Premier Degre
Porto Novo

Centre National de la Recherche Scientifique et Technique (G)
Ministere des Enseignements Technique et Superieur
Cotonou

Department des Recherches Agronomiques (G)
Ministere des Enseignements Technique et Superieur
Cotonou

Direction de l'Hydraulique (H)
Ministere des Travaux Publics, Mines, et Energie
B.P. 73
Cotonou

Direction de la Topographique (C)
Cadastre
B.P. 360
Cotonou

Office Béninois des Mines (G,R)
Ministère de l'Energie des Mines et de l'Hydrauliquepement
B. P. 249
Contonou
Phone: (229)31-03-09
Fax: (229)31-35-46

BHUTAN

Department of Geology and Mines (C,G,H)
Department of Trade, Industry and Power
P.O. Box 173
Thimphu
Head of the Department: Sonam Yangley
Phone: (975)2879
Telex: 215
(Answer back ID: MITI 215BT)

BOLIVIA

Centro de Investigaciones Geologicas (G)
Casilla 12198
Cota-Cota Calacoto
La Paz
Director: R. Santivanez Garcia
Phone: (591-2)793392/359581

Centro de Tecnologia Petrolera (G)
Casilla 727
Mamerto Cuellar s/n
Santa Cruz
Director: Mario Suarez R.

Instituto de Hidraulica e Hidrologia (H)
Casilla 699
La Paz
Director: Roger Matos Rueda
Phone: (591-2)795724

Instituto Geografico Militar (C)
Avenida Saavedra No. 2303
Cuartel Miraflores
La Paz
Director: Edgar Sandoval Calzadilla
Phone: (591-2)360513/378194
Fax: (591-2)368329

Instituto Minero Metalurgico (IMM) (G)
Casilla 600
Oruro
Director: Carlos Garron Ugarte

Obervatorio San Calixto (G)
Casilla de Correo 12656
La Paz
Director: Lawrence A. Drake
Phone: (591-2)356098
Fax: (591-2)376805
E-Mail: adrake@osc.bo

Servicio Nacional de Geologia y Mineria "SERGEOMIN" (C,G,H)
Federico Zuazo No. 1673
Casilla Postal 2729
La Paz
Executive Director: JoseAntonio Flores
Phone: (591-2)322022
Fax: (591-2)363474

Servicio Nacional de Meteorologia e Hidrologia (H)
Edificio La Urbana 6to. piso
Avenida Camacho 1485
La Paz

Director General: M. Canedo Daza
Phone: (591-2)355824

Yacimientos Petroliferos Fiscales Bolivianos (YPFB) (G)
Calle Bueno 185
Casilla 401
La Paz
General Manager: J. Flores Lopez
Phone: (591-2)328766
Telex: BX-5267

BOTSWANA

University of Botswana
Department of Geology
P.O. Box 00 22
Gaborone
Phone: (267) 351151
Fax: (267) 356591

Department of Mines
Ministry of Mineral Resources and Water Affairs
Private Bag 49
Gaborone

Department of Surveys and Lands (C)
Ministry of Local Government and Lands
Private Bag 37
Gaborone
Director: B.B.H. Morebodi
Phone: (267)353251
Telex: 2589 MLGL BD

Department of Water Affairs (H)
Ministry of Mineral Resources and Water Affairs
P.B. 29
Gaborone
Director: M. Sekwale
Phone: (267-31)351601
Telex: 2557 BD

Geological Survey Department (C,G,H)
Ministry of Mineral Resources and Water Affairs
Private Bag 14
Lobatse
Director: T.P. Machacha
Phone: (267-33)330428/330327
Fax: (267-33)332013
Telex: 2293 GEO BD

Ministry of Mineral Resources and Water Affairs
Department of Water Affairs
Private Bag 0029
Gaborone
B.B. Khupe
Phone: (267)3607100
Fax: (267)303508

BRAZIL

Association of Geoscientists for International Development
Instituto de Geociências
Universidade de São Paulo
Cidade Universitária
Caixa Postal 11.348
CEP 05422-970
São Paulo
Phone: (55-11) 818-4232
Fax: (55-11) 210-4958

Brazilian Geological Society
Instituto de Geociencias
Ciudade Universitaria
Caixa Postal 20897
CEP 01498
São Paulo
Phone: (55-11) 212-6166

Brazilian Geophysical Society
Sociedade Brasileira de Geofisica
Avenida Rio Branco, 156 sala 2510
20043-900 Rio de Janeiro - RJ
Marta Silvia Maria Mantovani
Phone: (55-21) 533 0064
Fax: (55-21) 533 0064
http://www.antares.com.br/sbgf/hpsbgf.htm

Companhia de Pesquisa de Recursos Minerais (CPRM) (G)
Servicio Geológico do Brasil
Av. Pasteur, 404 - Térreo
Praia Vermelha
22292-240 Rio de Janeiro, RJ
President: Carlos Oiti Berbert
Phone: (55-21)295-5337
Fax: (55-21)542-3647
http://www.cprm.gov.br

Departamento Nacional de Obras de Saneamento (DNOS) (H)
Rua Uruguaiana, 174 - 20 Andar
20091 Rio de Janeiro, RJ
Director General: P. Oscar Baier
Phone: (55-21)2337174/2538477
Telex: 22298

Departamento Nacional de Producao Mineral (DNPM) (G,R)
SAN - Q 1, Bloco B, 3 Andar

Sala 308
70040 Brasilia, D.F.
Director General: Miguel Navarrete Fernandez Jr.
Phone: (55-61)2242072/2247097
Fax: (55-61)2258274
Telex: 1116/1545

Departmento Nacional de Aguas e Energia Eletrica (DNAEE) (H)
Ministerio das Minas e Energia
SRTVS - Q. 701 - Edf. Palacio do Radio Bloco 3
70330 Brasilia, D.F.
Director General: Alvarino de A. Pereira
Phone: (55-61)2234592/2252617
Telex: 3765

Diretoria do Serviço Geográfico do Exército (DSG) (C)
SMU - QG Ex., Bloco F, 2 Pav.
70630-901
Brasília, D.F.
Director: João Venancio de Melo Neto
Phone: (55-061)415-4174
Fax: (55-061)415-5649

Fundação Instituto Brasileiro de Geografia e Estatistica (IBGE) (C,G)
Diretoria de Geociências (DGC)
Avenida Brasil 15671
Parada de Lucas
Rio de Janeiro - RJ
Director: Trento Natali Filho
Phone: (55-21)3911673
Fax: (55-21)4812650
http://www.ibge.gov.br

Mineral Resources Exploration Company
Companhia de Pesquisa de Recursos Minerais
Carlos Oiti Berbert
Av. Pasteur, 404
22.292-240-Urca
Rio de Janeiro
Phone: (55-21) 295-0032
Fax: (55-21) 542-3647

Ministry of Mines and Energy (MME) (G)
Esplanada dos Ministerios
Bloco J, 8 Andar
70056 Brasilia, D.F.
Minister: Vicente C. Fialho
Phone: (55-61)2237903
Telex: 1140

Petroleo Brasileiro S.A. (PETROBRAS) (G)
Avenida Republica do Chile, 65

24 Andar
20035-900 Rio de Janeiro, RJ
President: Joel Mendes Rennó
Phone: (55-21)534 1000
Fax: (55-21)240 9394
Telex: 22335

Projeto RADAM (C)
CRS 509, Bloco A, Loja 1 a 5
70360 Brasilia, D.F.
Director: David Almeida de Freitas
Phone: (55-61) 244-9432
Telex: 2243

Universidade de São Paulo - USP
Cidade Universitária - Instituto de Geociências
Rua do Lago, 562
Caixa Postal 20.899
CEP 05508-900 - São Paulo, SP
Phone: (55-11) 818-4244

Universidade Estadual de Campinas
Cidade Universitária Zeferino Vaz
Distrito de Barão Geraldo
Caixa Postal 6152
CEP 13081-970 - Campinas, SP
Phone: (0192) 39-1301

Universidade Estadual Paulista - UNESP
  Departamento de Geologia Sedimentar
  Av. 24-A, 1515, Bairro Bela Vista
  Caixa Postal 178
  CEP 13506-900 - Rio Claro, SP
  Phone: (0195) 34-0327
  Fax: (1095) 24-2445

  Instituto de Biociências
  Av. 24-A, 1515, Bairro Bela Vista
  Caixa Postal 199
  CEP 13506-900 - Rio Claro, SP
  Phone: (0195) 34-0244

  Instituto de Geociências e Ciências Exatas
  Rua 10, 2527, Bairro Santana
  Caixa Postal 178
  CEP 13500-230-Rio Claro-SP
  Phone: (0195) 34-0122
  Fax: (0195) 34-8250

BRUNEI DARUSSALAM

Petroleum Unit (R)
Exploration and Geology Division
Office of the Prime Minister
Bangunan Petroleum, Jalan Kebangsaan
Bandar Seri Begawan 2007

Division Head: Haji Harun bin Pengiran
  Abdul Rahman
Phone: (673-2)28932
Fax: (673-2)40270
Telex: BU2209

Survey Department of Brunei (C)
Ministry of Development
Bandar Seri Begawan 2070
Surveyor General: Awang Haji Yunos bin
  Mohd Noh
Phone: (673-2)43171
Telex: BU2228

BULGARIA

Bulgarian Academy of Sciences (G)
Geological Institute
Acad. G. Bonchev Street, Block 24
Sofia 1113
Director: Khrischo Khrischev
Phone: (359-2)723563
Fax: (359-2)724638
E-Mail: geoinst@bgearn acad.bg

Central Laboratory for Geodesy (C)
ul. 15 noemvri, 1
1040 Sofia
Director: Valentin Kotzev
Phone: (359-2)720841
Telex: 23381
E-Mail: kotzev@bgearn.acad.bg

**Central Laboratory for Space
  Research (G)**
6 Moskovska Street
Sofia 1000
Director: Dr. Boris Bonev
Phone: (359-2)801344
Telex: 23351

**Central Laboratory of Mineralogy and
  Crystallography (G)**
92 Rakovski Street
Sofia 1000
Director: Nikola Zidarov
Phone: (359-2)872450
Fax: (359-2)884979
E-Mail: mincryst@bgcict.acad.bg

**Committee of Geology and Mineral
  Resources (G)**
Bul. G. Dimitrov 22
Sofia 1000
President: Dr. Ivan Batandzhiev
Phone: (359-2)835260
Fax: (359-2)833568
Telex: 22502/22337 MIMEN AB

ENERGOPROEKT (G)
51, James Boucher Boulevbard
1407 Sofia
Managing Director: P. Mishev
Phone: (359-2)688160
Fax: (359-2)668951
E-Mail: office@enpro.bg

Geographic Institute (G)
Acad. G. Bonchev Street, Block 3
Sofia 1113
Director: Dr. Kirilo Mishev
Phone: (359-2)3361/700204
Telex: 22424

Geological Society of Bulgaria
Box 228
1000 Sofia
M. Staykov

Geophysical Institute (G)
Acad. G. Bonchev Street, Block 3
Sofia 1113
Director: G. Miloshev
Phone: (359-2)700264
Fax: (359-2)700226
Telex: 22632
E-Mail: geophys@bgearn.bitnet

Institute of Water Problems (H)
Acad. G. Bonchev Street, Block 1
Sofia 1113
Director: Ohanes Santurjian
Phone: (359-2)722572
Fax: (359-2)722577
E-Mail: santur@bgcict.acad.bg

**Laboratory for Geotechnology of
  Weak Earth (G)**
Foundations and Terrains
Acad. G. Bonchev Street, Block 24
Sofia 1113
Director: Prof. Minko Minkov
Phone: (359-2)3478/705361
Telex: 22424

**Laboratory of Seismic Mechanics and
  Earthquake Engineering (C)**
Acad. G. Bonchev Street, Block 3
Sofia 1113
Director: Dimitur Nenov
Phone: (359-2)720815
Fax: (359-2)700226
Telex: 22424

University of Mining and Geology (G)
1156 Sofia
Phone: (359-2)687224
Fax: (359-2)687512

Oceanography Institute (G)
Asparuhovo Quarter
P. Box 152
Varna 9000
Director: Dr. Zdravko Belberov
Phone: (359-52)775200

University of Sofia, Faculty of Geology and Geography (G)
15 Tzar Osvoboditel Bd.
Sofia 1000
Phone: (359-2)467057
Telex: 23926 SUKO BG

BURKINA FASO

Bureau of Mines and Geology of Burkina (G)
01 B.P. 601
Ouagadougou 01
Director: Jean-Léonard Compaore
Phone: (226)300227
Fax: (226)300187
Telex: 0978 5340 BUMIGEB BF

Burkina Geographic Institute (C)
B.P. 7054
Ouagadougou
Director: Baya Andre Bassole
Phone: (226)335916

Cadastral Direction (C)
B.P. 7054
Ouagadougou
Director: Dieudonne Bonanet

Interafrican Committee for Water Resources Studies (H)
B.P. 369
Ouagadougou
Director: Abdou Hassan
Phone: (226)307172

BURMA (see MYANMAR)

BURUNDI

Geographical Institute of Burundi (IGEBU) (C)
Ministry of Land Management, Tourism, and the Environment
B.P. 331 Bujumbura

Ministre: S.E.B. Sindaharaye
Cabinet Director: G. Ngendabanka

Ministry of Energy and Mines (G)
P. O. Box 745
Bujumbura
Director General of Geology and Mines: Mathias Sebahene
Phone: (257)22 22 78
Fax: (257)22 96 24
Telex: 5182 MEM BDI

CAMEROON

Department of Mines and Geology (C,E,G,H,R)
c/o Ministry of Mines, Water and Power
Yaounde
Director: Betah Sona Samuel
Phone: (237)23 34 04
Fax: (237)22 61 77
Telex: 8504 KN

Institute for Geological and Mineral Research (IRGM) (C,G,H)
B.P. 4110
Yaounde
Director: Ekodeck Georges Emmanuel
Phone: (237)21 03 16
Fax: (237)21 03 16
Chief Hydrology Service: Emmanuel Naah

Institute of Agronomic Research (C,H)
B.P. 2123
Yaounde
Director: Ayuk Takem
Phone: (237)233538

Ministry of Higher Education, Computer Services and Scientific Research (C,G,H)
B.P. 1457
Yaounde
Ministre: Babale Abdoulaye
Phone: (237)231650
Telex: 8414 KN

National Institute of Cartography (C)
P.O. Box 157
Avenue Monseigneur Vogt
Yaounde
Director: Paul Moby Etia
Phone: (237)22 29 21

CANADA

Alberta Department of Energy
Mineral Operations Division
Minerals Tenure Branch
9945 - 108 Street
Edmonton, Alberta T5K 2G6
David Coombs
Phone: (403) 427-7749
Fax: (403) 422-1123
E-Mail: coombsd@enr.gov.ab.ca

Alberta Geological Survey (G)
Alberta Energy and Utilities Board
6th Floor, Petroleum Plaza North
9945 - 108 Street
Edmonton, Alberta T5K 2G6
Jan Boon
Phone: (403) 422-1927
Fax: (403) 422-1459
E-Mail: boon@enr.gov.ab.ca
http://www.ags.enr.gov.ab.ca

Association of Exploration Geochemists
Box 523
Metropolitan Toronto
Rexdale, Ontario M9W 5L4
Arthur E. Soregaroli

Association of Professional Engineers, Geologists and Geophysicists of Alberta (APEGGA)
1500 Scotia Place
Tower 1
10060 Jasper Avenue
Edmonton, Alberta T5J 4A2
Robert Ross
Phone: (403) 426-3990
Fax: (403) 426-1877
http://www.apegga.com

Association of Professional Geologists and Geophysicists of Quebec
2549, boul. Rosemont
Bureau 101
Montreal, Quebec H1Y 1K5
Phone: (514) 725-0288
Fax: (514) 729-3380

Atlantic Geoscience Centre
Box 1006
Dartmouth, Nova Scotia B2Y 4A2
D. Prior
Phone: (902) 426-3448

Atlantic Geoscience Society
Geological Surveys Branch
Bathurst Regional Office
P.O. Box 50
Bathurst, New Brunswick E2A 3Z1
President: Michael Parkhill
Phone: (506) 547-2070
Fax: (506) 547-7694
E-Mail: gsbath@nbnet.nb.ca

British Columbia Geological Survey Branch
Ministry of Energy, Mines, and Petroleum Resources
1810 Blanshard St., 5th floor
Victoria, British Columbia V8V 1X4
W.R. Smyth
Phone: (604) 952-03724
Fax: (604) 952-0371.
E-Mail: rsmyth@galaxy.gov.bc.ca
http://natural.gov.bc.ca/

Canadian Continental Drilling Program
Geological Survey of Canada
Ottawa, Ontario K1A 0E8
Malcolm Drury
Phone: (613) 995-5478
Fax: (613) 996-9670
E-Mail: drury@gsc.emr.ca

Canadian Geological Survey
580 Booth St., 14th Floor
Ottawa, Ontario K1A 0E4
Assistant Deputy Minister: M.D. Everell
Phone: (613) 992-9983
Fax: (613) 996-8059
E-Mail: mdeverell@nrcan.gc.ca

Chief Geoscientist
601 Booth St., Room 209
Ottawa, Ontario K1A 0E8
J.M. Franklin
Phone: (613) 995-4482
Fax: (613) 996-8059
E-Mail: franklin@gsc.emr.ca
http://www.emr.ca/gsc/

Earth Science Information Centre
601 Booth St., Room 350
Ottawa, Ontario K1A OE8
B. Chen
Phone: (613) 995-4163
Fax: (613) 943-8742
E-Mail: bchen@gsc.emr.ca
http://ww.emr.ca/ess/esic/esic_e.html

Polar Continental Shelf Project
615 Booth Street, Room 233
Ottawa, Ontario K1A 0E9
Director: B. Hrycyk

Phone: (613) 947-1601
Fax: (613) 947-1611
http://www/emr.ca/gsc/pcsp/pcsp/html

Canadian Geological Survey — Minerals and Regional Geoscience Branch

601 Booth Street, Room 220
Ottawa, Ontario K18 0E8
J.M. Duke
Phone: (613) 995-4093
Fax: (613) 996-6575
E-Mail: mduke@gsc.emr.ca

Continental Geoscience Division
601 Booth Street, Room 459
Ottawa, Ontario K1A 0E8
J.E. King
Phone: (613) 995-4314
Fax: (613) 995-7322
E-Mail: jking@gsc.emr.ca

Geological Survey of Canada (Pacific)
Box 6000
Sidney, British Columbia V8L 4B2
S. Colvine
Phone: (604) 363-6438
Fax: (604) 363-6739
E-Mail: scolvine@gsc.emr.ca

Mineral Resources Division
601 Booth Street, Room 665
Ottawa, Ontario K1A 0E8
Director
Phone: (613) 996-9223
Fax: (613) 992-5694

Ottawa Subdivision
1 Observatory Crescent
Ottawa, Ontario K1A 0Y3
P. Basham
Phone: (613) 995-0904
Fax: (613) 992-8836
E-Mail: basham@seismo.emr.ca

Vancouver Subdivision
100 West Pender
Vancouver, British Columbia V6B 1R8
C. Hickson
Phone: (604) 666-0529
Fax: (604) 666-1124
E-Mail: chickson@gsc.emr.ca

Canadian Geological Survey — Sedimentary and Marine Geoscience Branch

601 Booth Street, Room 216
Ottawa, Ontario K1A 0E8
R.T. Haworth
Phone: (613) 995-2340
Fax: (613) 996-6575
E-Mail: haworth@gsc.emr.ca

Geological Survey of Canada (Atlantic)
Box 1008
Dartmouth, Nova Scotia B2Y 4A2
Director
Phone: (902) 426-3448
Fax: (902) 426-1466

Geological Survey of Canada (Calgary)
3303 - 33 Street, N.W.
Calgary, Alberta T2L 2A7
G.D. Mossop
Phone: (403) 292-7049
Fax: (403) 292-5377
E-Mail: mossop@gsc.emr.ca

Quebec Geoscience Centre
2535 boul. Laurier
C.P. 7500
Ste. Foy, Québec G1V 4C7
A. Achab
Phone: (418) 654-2604
Fax: (418) 654-2615
E-Mail: achab@gsc.emr.ca

Terrain Sciences Division
601 Booth Street, Room 361
Ottawa, Ontario K1A 0E8
J-S. Vincent
Phone: (613) 995-4938
Fax: (613) 992-0190
E-Mail: svincent@gsc.emr.ca

Canadian Geophysical Union

Geological Survey of Canada
1 Observatory Crescent
Ottawa, Ontario K1A OY3
David W. Eaton
Phone: (613) 947-2783
Fax: (613) 992-8836
E-Mail: eaton@cg.emr.ca

Canadian Geoscience Council

Department of Earth Sciences
University of Waterloo
Waterloo, Ontario N2L 3G1
Mario Coniglio
Phone: (519) 885-1211, EXT 2066
Fax: (519) 746-7484
E-Mail: coniglio@sciborg.uwaterloo.ca

Canadian Geotechnical Society

501-170 Attwell Drive
Rexdale, Ontario M9W 5Z5
A.G. Stermac
Phone: (416) 674-0366
Fax: (416) 674-9507
http://www.inforamp.net/~cgs

Canadian Institute of Mining, Metallurgy, and Petroleum

Suite 1210

3400 de Maisonneuve Boulevard West
Montreal, Quebec H3Z 3B8
Phone: (514) 939-2710
Fax: (514) 939-2714
E-Mail: publications@asinet.net

Canadian Quaternary Association (E,G,H)

Geological Survey Branch
Mineral Resources Division
Department of Natural Resources and Energy
Box 6000
Fredericton, New Brunswick E3B 5H1
A.G. Pronk
Phone: (506) 453-2206
Fax: (506) 444-4176
E-Mail: tgpronk@gov.nb.ca

Canadian Society of Petroleum Geologists (G)

Suite 505
206 Seventh Ave. S.W.
Calgary, Alberta T2P 0W7
Tim Howard
Phone: (403) 264-5610
Fax: (403) 264-5898
http://www.cspg.org

Canadian Society of Soil Science

5320 122 St.
Edmonton, Alberta T6H 3S5
Yash P. Kalra
Phone: (403) 435-7210
Fax: (403) 435-7359
E-Mail: ykalra@nofc.forestry.ca

Canadian Well Logging Society

Suite 229
640 Fifth Ave. S.W.
Calgary, Alberta T2P 0M6
Sandra Scott
Phone: (403) 269-9366
Fax: (403) 269-2787

Central Survey and Mapping Agency (C)

2nd Floor, 2151 Scarth Street
Regina, Saskatchewan S4P 3V7
Director of Geodata: George W. Meggitt
Phone: (306)787-4880
Fax: (306)787-4617

Club de Mineralogie de Montreal Inc.

C.P. 305
Succursale St-Michel
Montreal, Quebec H2A 3M1
Andre Brisebois
Phone: (514) 729-6416

Continental Geoscience Division

Room 459
601 Booth St.
Ottawa, Ontario K1A 0E8
J.E. King
Phone: (613) 995-4314
E-Mail: jking@gsc.emr.ca

Coordination and Planning Division

Room 213
601 Booth St.
Ottawa, Ontario K1A 0E8
A.G. Plant
Phone: (613) 995-9495

Department of Energy, Mines, and Petroleum Resources

Government of the Northwest Territories
Box 1320
Yellowknife, Northwest Territories X1A 2L9
Michael Cunningham
Phone: (403) 920-3217
Fax: (403) 873-0254
E-Mail: michael@inukshuk.gov.nt.ca

Department of Indian and Northern Affairs Canada

Mineral Resources Directorate, Geology Division
P.O. Box 1500
Yellowknife, Northwest Territories X1A 2R3
W.A. Padgham
Phone: (403) 669-2635
Fax: (403) 669-2709
E-Mail: padghamw@inac.gc.ca

Department of Mines and Energy

Government of Newfoundland and Labrador
Box 8700
St. John's, Newfoundland A1B 4J6
C. Patey
Phone: (709) 729-3159
Fax: (709) 729-4491
http://www.geosurv.gov.nf.ca

Economic Development and Tourism

Energy and Minerals Section
Box 2000
Charlottetown, Prince Edward Island C1A 7N8
Wayne MacQuarrie
Phone: (902) 368-5010
Fax: (902) 368-5010
E-Mail: rgestabrooks@gov.pe.ca

Ecosystem and Environmental Resources Directorate (E)

Knowledge Integration Division
Department of the Environment
Place Vincent Massey

351 St. Joseph Boulevard
Hull, Quebec K1A 0H3
Chief: Art Goldsmith
Phone: (819)953-1547
Fax: (819)997-3822
E-Mail: goldsmia@cpits1.am.doe.ca

Geological Association of Canada

Cordilleran Section
Box 398
Station A
Vancouver, British Columbia V6C 2N2
Greg Dipple
http://nereus.geology.ubc.ca/gac/gac.html

Geological Association of Canada

Department of Earth Sciences
Memorial University of Newfoundland
St. John's, Newfoundland A1B 3X5
R.N. Hiscott
Phone: (709) 737-7660
Fax: (709) 737-2532
http://www.esd.mun.ca/~gac

Geological Survey of Newfoundland and Labrador

Department of Mines and Energy
Government of Newfoundland and Labrador
Box 8700
St. John's, Newfoundland A1B 4J6
Catherine Patey
Phone: (709) 729-3159
Fax: (709) 729-3493
http://www.geosurv.gov.nf.ca

Inter-Union Commission on the Lithosphere

c/o Geological Survey of Canada
Room 186
601 Booth St.
Ottawa, Ontario K1A 0E8
M.J. Berry
Phone: (613) 995-5484
Fax: (613) 952-9088
E-Mail: berry@gsc.emr.ca

International Association of Geochemistry and Cosmochemistry

Applied Geoscience Branch
AECL Research
Pinawa, Manitoba R0E 1L0
Mel Gascoyne
Phone: (204) 753-2311
Fax: (204) 753-2703
http://www.ent.msu.edu/~long/iagc

International Association of Hydrogeologists

Canadian National Chapter

1111 Lake Wapta Place S.E.
Calgary, Alberta T2J 2P4
Laurra Olmsted
Phone: (408) 297-3600
Fax: (403) 236-7033
E-Mail: olmsted@eba.ca

International Association of Hydrological Sciences

Department of Geography
Wilfrid Laurier University
Waterloo, Ontario N21 3C5
Secretary General: Gordon J. Young
Phone: (519) 884-1970 EXT. 2387
Fax: (519) 846-0968
E-Mail: 44iahs@machl.wlu.ca

International Association on the Genesis of Ore Deposits

Geological Survey of Canada
601 Booth St.
Ottawa, Ontario K1A 0E8
Ian R. Jonasson
Phone: (613) 996-2766
Fax: (613) 943-1286
Telex: 053-3117
http://www.emr.ca/gsc/gied/cgic/catalog.html

Manitoba Energy and Mines

Geological Services Branch, Minerals Division
1395 Ellice Ave., Suite 360
Winnipeg, Manitoba R3G 3P2
W.D. McRitchie
Phone: (204) 945-6559
Fax: (204) 945-1406

Mineralogical Association of Canada

P.O. Box 78087
Meriline Postal Outlet
1460 Merivale Road
Ottawa, Ontario K2E 1B1
Phone: (613) 226-4651
Fax: (613) 226-4651
E-Mail: canmin.mac.ottawa@sympatico.ca

Mining Association of British Columbia

Box 12540
860-1066 West Hastings St.
Vancouver, British Columbia V6E 3X1
Gary K. Livingstone
Phone: (604) 681-4321
Fax: (604) 681-5305

National Water Research Institute

Box 5050
Burlington, Ontario L7R 4A6
R.J. Daley

Phone: (905) 336-4625
Fax: (905) 336-4989

Natural Resources Canada (C,G)

Earth Sciences Sector
Business Development
615 Booth Street, Room 401
Ottawa, Ontario K1A 0E9
Phone: (613)995-0314
Fax: (613)927-2189
E-Mail: info@geocan.nrcan.gc.ca

New Brunswick Minerals and Energy Division

Geological Surveys Branch
Department of Natural Resources and Energy
Box 6000
Fredericton, New Brunswick E3B 5H1
J.L. Davies
Phone: (506) 453-2206
Fax: (506) 453-3671

North American Commission on Stratigraphic Nomenclature

Precambrian Geoscience Section
Ontario Geological Survey
933 Ramsery Lake Road
Sudbury, Ontario P3E 6B5
R. Michael Easton
Phone: (705) 670-5985
Fax: (705) 690-5953

Northwest Territories Geology Division

Geology Division
Box 1500
Yellowknife, Northwest Territories X1A 2R3
W.A. Padgham
Phone: (403) 669-2635
Fax: (403) 669-2709

Nova Scotia Department of Natural Resources

Box 689
Halifax, Nova Scotia B3J 2T9
Scott Swinden
Phone: (902) 424-8135
Fax: (902) 424-7735
E-Mail: hsswinde@gov.ns.ca

Nova Scotia Department of Natural Resources

Box 689
Halifax, Nova Scotia B3J 2T9
Library
Phone: (902) 424-8633
Fax: (902) 424-7735
E-Mail: nsdnrlib@gov.ns.ca

Ontario Geological Survey

933 Ramsey Lake Road
Sudbury, Ontario P3E 6B5
Cameron Baker
Phone: (705) 670-5904
Fax: (705) 670-5905
E-Mail: baker_c@torv05.ndm.gov.on.ca

Ontario Mining Association

Suite 1501
110 Yonge St.
Toronto, Ontario M5C 1T4
Patrick Reid
Phone: (416) 364-9301
Fax: (416) 364-5986

Prince Edward Island Department of Energy and Minerals

Box 2000
Charlottetown, Prince Edward Island C1A 7N9
Wayne MacQuarrie
Phone: (902) 368-5025
Fax: (902) 368-5544

Quebec Ministre des Ressources Naturelles (G)

5700 4e Avenue Ouest
bur. A-208
Charlesbourg, Quebec G1H 6R1
Jean Louis Caty
Phone: (418) 643-5159
Fax: (418) 643-2816

Saskatchewan Energy and Mines

Exploration and Geological Services
1914 Hamilton St.
Regina, Saskatchewan S4P 4V4
G.C. Patterson
Phone: (306) 787-2476
Fax: (306) 787-7338
E-Mail: patgc@mailhost.sasknet.sk.ca

Saskatchewan Geological Society

Box 234
Regina, Saskatchewan S4P 2Z6
Robert Troyer

Toronto Geological Discussion Group

Richardson Greenshields of Canada Ltd.
1400-130 Adelaide St. West
Toronto, Ontario M5H 1T8
R. Goldie

Yukon Territory Exploration and Geological Services Division

Northern Affairs Program
345-300 Main St.
Whitehorse, Yukon Y1A 2B5
Trevor Bremner

Phone: (403) 667-3201
Fax: (403) 667-3198
E-Mail: bremnert@inac.gc.ca

CENTRAL AFRICAN REPUBLIC

High Commission for Scientific and Technological Research (C,G,H)
Bangui
High Commissioner: Gaston Mandata-Nguerekata
Phone: (236)612107

Institut Geographique National (C)
B.P. 165
Bangui
Directeur: M. Xavier Barillot

Ministry of Energy, Mines, Geology and Hydraulics (G,H)
Bangui
Minister: Michel Salle
Phone: (236)612944/613944

Ministry of Public Works and Town Development (C,G)
Bangui
Minister: Jacques Kitte
Phone: (236)614590

Ministry of Rural Development (G)
Bangui
Minister: Jean Willybiro-Sacko

Ministry of Water, Lands, Game Fishing, and Tourism (G)
Bangui
Minister: Raymond Mbitikon
Phone: (236)611444

CHAD

Directorate of Geological and Mining Research (D.R.G.M.) (C,G,R)
B.P. 816
N'Djamena
Director: Salibou Garba
Phone: (235)512630
Fax: (235)512565

Documentation Centre
Head of the Centre: Taglo Jaques
Phone: (235)512936
Fax: (235)516330

Geochemical Lab
Chief Chemist: N'Diguina Ndilnger Torbaye
Phone: (235)512544
Fax: (235)516330

Geology Department
Head of the Department: N'djekounde Moy Sey
Phone: (235)515482
Fax: (235)512565

Mines Department
Head of the Department: Oumar Mahamat Abba
Phone: (235)515482
Fax: (235)512565

Directorate of Hydraulics and Water Clean Up (DHA) (H,R)
B.P. 816
N'Djamena
Director: Moussa Moustapha Terap
Phone: (235)513437
Fax: (235)512565

Directorate of Petroleum, New and Renewable Energy (DPENR) (E,G,R)
B.P. 816
N'Djamena
Director: Yolla Aguenade Zongre
Phone: (235)513850
Fax: (235)512565

Department of New and Renewable Energy
Head of the Department: Mahamat Bourdjo
Phone: (235)512566
Fax: (235)512565
Head of the Department: Oumar Matar Breme
Phone: (235)513437
Fax: (235)512565

Petroleum Department
Deputy Director: Abdel-Hamid Mahamat Ali
Phone: (235)513437
Fax: (235)512565

National Pastoral and Village Hydraulics Office (ONHPV) (G,H)
B.P. 750
N'Djamena
Director: Adoum Diar
Phone: (235)515994/516091
Fax: (235)512565

CHILE

Centro de Investigaciones Minero-Metalurgicas (G)
Parque Institucional 6500, Lo Curro
Santiago
Director: Werner Sechleim
Phone: (56-2)2289544

Chilean Association of Seismology and Earthquake Engineering
Associacion Chilena de Sismologia e Ingenieria Antisismica
Casilla 2796
Santiago

Empresa Nacional del Petróleo (ENAP) (G)
Unidad Contratos de Operación
Ahumada 341, 8 Piso
Santiago
Chief: Eduardo Gonzales
Phone: (56-2)6727892
Fax: (56-2)6380164

Geological Society of Chile
Sociedad Geologica de Chile
Valentin Letelier 20 Depto. 401
Casilla 13667
Correo
Santiago
Fernando Henriquez
Phone: (56-2) 698-0481

Instituto de Investigacion de Recuros Naturales (IREN) (G)
Corporacion de Fomento de la Produccion
Manuel Montt 1164
Santiago
Executive Director: E. Junemann
Phone: (56-2)2236641

Instituto Geográfico Militar (C)
Nueva Santa Isabel 1640
Santiago
Director: Enrique Gillmore Callejas
Phone: (56-2) 6968221
Fax: (56-2) 6988278
E-Mail: igm@reuna.cl

National Committee of Geography, Geodesy and Geophysics
Comite Nacional de Geografia, Geodesia y Geofisica
Nueva Santa Isabel 1640
Santiago

National Society of Mining
Sociedad Nacional de Mineria
Teatinos No. 20, Of. 33
Casilla 1807
Santiago

Servicio Aerofotograметrico (C)
Casilla 67
Correo Los Cerrillos
Santiago
Director: Comandante de Grupo del Aire Ricardo Contreras
Phone: (56-2)573234

Servicio Hidrografico y Oceanografico de la Armada (C,H)
Errázuriz 232
Playa Ancha
Casilla 324
Valparaíso
Director: Hugo Gorziglia A.
Phone: (56-32)282697
Fax: (56-32)283537
E-Mail: shoa@huelen.reuna.cl

Servicio Nacional de Geologia y Mineria (SERNAGEOMIN) (G)
Avenida Santa María 0104
Providencia
National Director: Ricardo Troncoso San Martin
Phone: (56-2)7375050
Fax: (56-2)7372026
E-Mail: sernageo@reuna.cl
Sub-Director of Geology: Jorge Skarmeta
Phone: (56-2)375050
Sub-Director of Mines: Sergio Schindler
Phone: (56-2)6985103

CHINA

Academia Sinica (Chinese Academy of Sciences) (C,E,G,H)

Changchun Institute of Geography
10 Gongnong Dalu
Changchun 130021
Jilin Province
Director: Liu Zheming
Phone: (86)52620

Changsha Institute of Geotectonics
Tongzipo
Changsha, Hunan Province 410013
Zhao Sheng-cai
Phone: (86-0731) 8859137
Fax: (86-0731) 8859137

Division of Earth Sciences
Salihe Lu No. 52
Beijing 100863

23

Director: Tu Guangzhi
Phone: (86-1)863805
Telex: 22474 ASCHI OCN

Guangzhou Institute of Geochemistry
Wushan
Guangzhou 510640
Chair: Zhao Zhen-hua
Phone: (86-020) 5519755
Fax: (86-020) 5514130

Institute of Desert Research
14 West Donggang Lu
Lanzhou 730000
Gansu Province
Director: Zhu Zhenda
Phone: (86)226725

Institute of Geochemistry
73, Guanshui Road
Guiyang, Guizhou Province 550002
Chair: Xie Hong-shen
Phone: (86-0851) 5820906
Fax: (86-0851) 5822982
Telex: 7181

Institute of Geochemistry
73 Guanshui Road
Guiyang 550002
Guizhou Province
Director: Xie Hongsen
Phone: (86-851)5814757
Fax: (86-851)5822982
E-Mail: gas@maple.edu.cn

Institute of Geography
Building 917, Datun Road, Anwai
Beijing 100101
Director: Liu Yanhua
Phone: (86-10)64914841
Fax: (86-10)64911844
E-Mail: zhengd@sun.ihep.ac.cn

Institute of Geology
Qijiahuozi, Desheng Menwai
P.O. Box 634
Beijing 100011
Director: Wang Sijing
Phone: (86-1)446551, Ext. 315

Institute of Geology and Paleontology
39 Beijing Donglu
Nanjing 210000
Director: Wu Wangshi
Phone: (86)631400

Institute of Geophysics
Qinghua Donglu
Haidian District
Beijing 100083
Director: Chen Zongji
Phone: (86-1)2012497

Institute of Geotectonics
Hexi Tongzipo
Changsha 410013
Hunan Province
Director: Liu shunsheng
Phone: (86-731)8859137
Fax: (86-731)8859137

Institute of Glaciology and Cryopedology
14 Donggang Xilu
Lanzhou 730000
Gansu Province
Director: Xie Zichu
Phone: (86)21894

Institute of Mountain Hazards and Environment
No. 9 Block 4, Renmin Road
P.O. Box 417, Chengdu
Sichuan 610041
Director: Zhong Xianghao
Phone: (86-28)5552258
E-Mail: clcas@rose.cnc.ac.cn

Institute of Oceanology
7 Nan-Hai Road
Qingdao 266071
Deputy Director: Naisheng Li
Phone: (86-532)287-9062
Fax: (86-532)287-0882

Institute of Pedology
71 Beijing Donglu
Nanjing 210008
Jiangsu Province
Director: Tu Qingying
Phone: (86)713781

Institute of Remote Sensing Application
Bldg. 917, Beishatan
Deshengmenwai
Beijing 100012
Director: Tong Qingxi
Phone: (86-1)449382

Institute of Rock and Soil Mechanics
Xiaohongshan, Wuchang
Wuhan 430071
Hubei Province
Director: Bai Shiwei
Phone: (27) 7869251
Fax: (27) 7863386
E-Mail: swbai@dell.whrsm.ac.cn

Institute of Salt Lakes
7 Xinning Lu
Xining 810000
Qinhai Province
Director: Zhang Pengxi
Phone: (86)55135

Institute of Soil Sciences
71, East Beijing Road
Nanjing 210008

Zhao Qi-guo
Phone: (86-025) 7712572
Fax: (86-025) 3353590
E-Mail: jiangsu@chinamai/sh.sprinl.com

Institute of Surveying and Geophysics
54 Xudong Lu
Wuchang, Wuhan 430077
Hebei Province
Director: Xu Houze
Phone: (86)813401

Institute of Vertebrate Paleontology and Paleoanthropology
Xizhimenwai Street 142
Beijing 100044
Chair: Qiu Zhan-xiang
Phone: (86-010) 8354669
Fax: (86-010) 8312683

Institute of Vertebrate Paleontology of Paleoanthropology
142 Xizhimenwai Street
Beijing 100044
Director: Qiu Zhuding
Phone: (86-10)68354669
Fax: (86-10)68337001

Lanzhou Institute of Geology
West Donggang Road, No. 198
Lanzhou 730000
Chair: Wang Xian-bin
Phone: (86-0931) 8827981
Fax: (86-0931) 8418667

Lanzhou Institute of Geology
198 Donggang Xilu
Lanzhou 730000
Gansu Province
Director: Luo Binjie
Phone: (86)27981

Nanjing Institute of Geology and Paleontology
East Beijing Road, No. 39
Nanjing 210008
Chair: Cao Rui-ji
Phone: (86-025) 7713534
Fax: (86-025) 3357026
Telex: 342301 NJIGP CN

Northwest Institute of Soil and Water Conservation
Yangling Town Station
Xianyang 712100
Shaanxi Province
Director: Yang Wenzhi

South China Sea Institute of Oceanology
164 West Xingang Lu
Guangzhou 510301
Guangdong Province
Director: Pan Jinpei

Phone: (86-20)84451335
E-Mail: itnhlib@scut.edu.cn

Xinjiang Institute of Biology, Pedology and Desert Research
No. 40-3 South of Beijing Road
Urumqi
Xinjiang Uygur Autonomous Region
Director: Li Chongshun
Phone: (0086-991)3839133
Fax: (0086-991)3835459

Xinjiang Institute of Geography
40 South Beijing Street
Urumqi 830011
Xinjiang Uygur Autonomous Region
Director: Han Delin
Phone: (86-91)3837482
Fax: (86-91)3835459
Telex: 77152 KWTWZ CN

Beijing Natural History Museum
No. 126, Tian Qiao Street
Beijing 100050
Phone: (86-10) 6702 4431-3052
Fax: (86-10) 6702 1254

Beijing Research Institute of Uranium Geology (G)
Xiaoguan Dongjie No. 10
Anding Menwai
P.O. Box 9818
Beijing 100029
Director: Zhao Fengmin
Phone: (86-10)64914830
Fax: (86-10)64917143

Beijing University of Science and Technology
Resources Engineering College
30, Xue Yuan Lu, Hai Dian
Beijing 100083
Chair: Ni Wen
Phone: (86-010) 2019944-2465
Fax: (86-010) 2017283

Bureau of Geology and Mining
Liupukang, Andingmenwai, Xicheng District
Beijing 100011
Phone: (86-10) 6201 9933-304
Fax: (86-10) 6204 4386

Bureau of Petroleum and Marine Geology
Building 7, Honglian Beicun, Hadian District
Beijing 100083
Phone: (86-10) 6225 5533-3082
Fax: (86-10) 6225 1536

Central South University of Technology
Department of Geology
Zuojialong
Changsha, Hunan Province 410083
Chair: Peng You-lin
Phone: (86-0731) 8883111
Fax: (86-0731) 8851136

Chengdu Institute of Technology
No. 1, Erxianqiao Dongsanlu
Chengdu, Sichuan Province 610059
Chair: He Zhen-hua
Phone: (86-028) 3334712
Fax: (86-028) 3334963

China Association of Geological Education
No. 29, Xueyuanlu, Haidian District
Beijing 100083
Phone: (86-10) 6202 2244-3162
Fax: (86-10) 6614 5494

China University of Geosciences
29, Xueyuan Road, Hai Dian District
Beijing 100083
Chair: Zhao Peng-da
Phone: (86-010) 2017228
Fax: (86-010) 2014874

China University of Geosciences, Wuhan
Yujiashan
Wuhan, Hubei Province 430074
Chair: Zhao Peng-da
Phone: (86-027) 7801330
Fax: (86-027) 7801763

China Institute of Hydrogeology and Engineering Geology Exploration
20 Dahuisi, Hai Dian
Beijing 100081
Chair: Wu Jing-yang
Phone: (86-010) 8350261
Fax: (86-010) 8311627

China University of Mining and Technology
College of Resources and Environmental Science
Xuzhou, Jiangsu Province 221008
Chair: Liu Huan-jie
Phone: (86-0516) 3888653-638
Fax: (86-0516) 3888682

China University of Mining and Technology
Geology Department
Beijing Graduate School
D11, Xue Yuan Road
Beijing 100083
Chair: Guo Ying-ting
Phone: (86-010) 2017641-248
Fax: (86-010) 2025016

China National Coal Corporation (G)
21 Heping Beilu
P.O. Box 1409
Beijing 1000713
Manager General: Yu Hongen
Phone: (86-1)4214591
Telex: 22660 CNMCI CN

Central Coal Mining Research Institute, Xi'an Branch
44, Yanta Road (N)n
Xi'an 710054
Shaanxi Province
Director: Pan Zhenwu
Phone: (86-29) 7214117
Fax: (86-29) 7234674

China National Geological Exploration Centre of Building Materials Industry
No. 11, Beishunchengjie, Xinei
Beijing 100035
Phone: (86-10) 6225 3094
Fax: (86-10) 6225 3105

China National Nonferrous Metals Industry Corporation
Beijing Institute of Geology for Mineral Resources
Anwai, Beiyuan
Beijing 100012
Cahir: Chen Zheng-jie
Phone: (86-010) 4232233
Fax: (86-010) 4232384

China National Nuclear Corporation
Bureau of Geology
P.O. Box 762
Beijing 100013
Phone: (86-10) 6420 1122-292
Fax: (86-10) 6420 1122-286

China National Offshore Corporation
P.O. Box 4705
Beijing 100027
Phone: (86-10) 6466 2989
Fax: (86-10) 6466 2994

China National Petroleum Corp. (G)
P.O. Box 776
Beijing 100724
President: Wang Tao
Phone: (86-10)62016107
Fax: (86-10)62094806

China Ocean Mineral Resources Association
No. 1, Fu Xing Men Wai Street
Beijing 100860
Phone: (86-10) 6853 2211-7595
Fax: (86-10) 6853 3318

Chinese Academy of Geoexploration
No. 23, Xueyuanlu, Haidian District
Beijing 100083
Phone: (86-10) 6201 8811-344
Fax: (86-10) 6617 6612

Chinese Academy of Geological Sciences (CAGS) (C,E,G,H)
26 Baiwanzhuang Road
Beijing 100037
President: Chen Yuchuan
Phone: (86-10)6835-1928
Fax: (86-10)6831-0894
E-Mail: cagsdic@public.bta.net.cn

Institute of Gemechanics
Fahuasi, Xijiao
Beijing 100081
Director: Wu Ganguo
Phone: (86-10)68412309
E-Mail: yangzy@public3.bta.net.cn

Institute of Geology
Director: Guo Yunling
Phone: (86-1)8311546

Institute of Geomechanics
11, Minzuxueyuan Nanlu, Hai Dian District
Beijing 100081
Chair: Wu Gan-guo
Phone: (86-010) 8422326
Fax: (86-010) 8422326

Institute of Geophysics
A-11, Datun Road, Chao Yang District
Beijing 100101
Chair: Xu Wen-yao
Phone: (86-010) 2011118
Fax: (86-010) 2031995
E-Mail: huangw@bepc2.ihep.ac.cn

Institute of Mineral Deposits
26, Baiwanzhuang Road
Beijing 100037
Chair: Hu Yun-zhong
Phone: (86-010) 8323294
Fax: (86-010) 8310894
Telex: 222721 CAGS CN

Institute of Mineral Deposits
Director: Yu Zhihong
Phone: (86-1)8312403

Shenyang Institute of Geology and Mineral Resources
No. 25, Beijing Street, Huanggutun District
Shenyang, Liaoning Province 110032
Chair: Gu Feng
Phone: (86-024) 6847571
Fax: (86-024) 6843124
Telex: 6005

Tianjin Institute of Geology and Mineral Resources
No. 4, 8th Road, Dazhigu
Tanjin 300170
Chair: Shen Bao-feng
Phone: (86-022) 4314292
Fax: (86-022) 4314292

Xi'an Institute of Geology and Mineral Resources
166, East Youyi Road
Xi'an 710054
Chair: Zhu Sheng-ying
Phone: (86-029) 5251266
Fax: (86-029) 5251266

Chinese Academy of Surveying and Mapping
16 Beitaiping Road, Haidian District
Beijing 100039
Phone: (86-10)6821-2277 EXT. 284
Fax: (86-10)6821-8654

Chinese Committee of Geothermal Exploitation Management
No. 4, Nanwei Road, Xuan Wu District
Beijing 100050
Phone: (86-10) 6301 4394

Chinese Institute of Geology and Mineral Resources Information
277 Fuwai North Street
Beijing 100037
Chair: Li Yu-wei
Phone: (86-010) 8352874
Fax: (86-010) 8323270

Chinese National Nuclear Industry Company
Beijing Research Institute of Uranium Geology
No. 10, Xiao Guan Dong Li, Chaoyang District
Beijing 100029
Chair: ZXhaoi Feng-min
Phone: (86-101) 4912211
Fax: (86-010) 4917143

Dept and Graduate Inst of Geology
National Taiwan University

245 Choushan Road
Taipei 106-17, Taiwan
People's Republic of China
Hsien Ho Tsien
Phone: 3630231, ext
Fax: 2-3636095

General Bureau of Geology and Exploration, CNNC

Dayangfang An Wa
Beijing 100012
Phone: (86-10) 6423 2233-424
Fax: (86-10) 6423 2384

Geological Bureau of Ministry of Metallurgical Industry

No. 46, Dongsixi Dajie
Beijing 100711
Phone: (86-10) 6513 3322-4233
Fax: (86-10) 6513 3814

Gold Bureau, Ministry of Metallurgical Industry

Qingnianhu
Beijing 100011
Phone: (86-10) 6426 9337
Fax: (86-10) 6426 3355

Gold Headquarters, Ministry of Metallurgical Industry

Langfang
Hebei 102800
Phone: (86-10) 6423 1122-517
Fax: (86-10) 6423 2373

Hebei College of Geology

No. 40, Jian Hua Nandajie
Shijiazhuang 050031
Chair: Yang Chang-ming
Phone: (86-0311) 5057701-2067
Fax: (86-0311) 5052562

Hunan Bureau of Geology and Mineral Exploration and Development

No. 67, Renmin Road, Changsha
Hunan 410011
Phone: (731) 553 5011-6007

Institute of Geology (G)

Chinese Academy of Geological Sciences
26 Baiwanzhuang Road
Beijing 100037
Director: Xu Zhiqin
Phone: (86-10)68329504
Fax: (86-10)68311293
E-Mail: yangjsui@public.bta.net.cn

Institute of Hydrogeology and Engineering Geology (G,H)

Bureau of Geology and Mineral Resources of Hebei Province
76 Gongnong Road, Shijiazhuang City
Hebei Province

Chengdu Institute of Geology and Mineral Resources

82/3, No. 1 Ring Road (N)
Chengdu, Sichuan 610082
Director: Liu Baojun
Phone: (86-28)3332657
Fax: (86-28)3332657
E-Mail: clbj@shell.scsti.ac.cn

Institute of Geology and Mineral Resources

21 Gangyao Lu, Yichang
Hebei Province
Director: Tan Zhongfu
Phone: (86-311)21635

Institute of Geology and Mineral Resources

4 Road, Dazhigu
Tianjin City
Director: Shen Baofeng
Phone: (86 22)4314386
Fax: (86 22)4314292

Institute of Geology and Mineral Resources

Beiling Street, Shenyang
Liaoning Province
Director: Gu Feng
Phone: (86)67571/67572

Institute of Geology and Mineral Resources (Volcanic Rocks)

534 Zhongshan Donglue
Nanjing
Jiangsu Province
Director: Lu Zhigang
Phone: (86)643992

Institute of Karst

Qixing Lu, Guilin
Guangxi
Director: Xu Zhenxin
Phone: (86)445151

Institute of Multipurpose Utilization of Mineral Resources of MGMR

5.3 Section of South Second Round-City Road
Chengdu, Sichuan Province 610041
Director: Ding Qiguang
Phone: (028)5551110
Fax: (028)5551383

Institute of Rock and Mineral Analysis

26 Baiwanzhuang, Dajie
Beijing 100037

Director: Li Jiaxi
Phone: (86-10)68336788
Fax: (86-10)68320365

Xi'an Institute of Geology and Mineral Resources
166 Youyi Road
Xi'an 710054
Shaanxi Province
Phone: (86)029-5251266
Fax: (86)029-7802701

Zhengding
Hebei Province, 050803
Director: Ren Fuhong
Phone: (086-311)8022028
Fax: (086-311)8021225

Institute of Rock and Mineral Analysis
26, Baiwanzhuang Street
Beijing 100037
Chair: Li Jia-xi
Phone: (86-010) 8323635
Fax: (86-010) 8310894
Telex: 222721 CAGS CN

Ministry of Geology and Mineral Resources (G)
64 Fucheng Mennei Street
Beijing 00812
Minister: Zhue Xun
Phone: (86-1)664845
Telex: 22531 MGMRC CN

Chengdu Institute of Geology and Mineral Resources
82/3, No. 1 Road (North)
Chengdu, Sichuan Province 610082
Chair: Liu Bao-jun
Phone: (86-028) 3335030
Fax: (86-028) 3332657

Institute of Hydrogeology and Engineering Geology
Zhengding, Hebei Province 050803
Chair: Fei Jin
Phone: (86-0311) 8021225
Fax: (86-0311) 8021225
E-Mail: feij@sun.ihep.ac.cn

Institute of Marine Geology
P.O. Box 18
Qingdao, Shandong Province 266071
Chair: Liu Shou-quan
Phone: (86-0532) 5814651-2207
Fax: (86-0532) 5810533

Institute of Petroleum Geology
31, College Road, Haidian District
Beijing 100083
Chair: Wang Ting-bin
Phone: (86-010) 2042233-438
Fax: (86-101) 2024674

Ministry of Metallurgical Industry
Tianjin Geological Academy
42, Youyi Road
Tianjin 300061
Chair: Hou Zong-lin
Phone: (86-022) 8357460
Fax: (86-022) 8357460

Ministry of Water Resources (H)
Baiguanglu, Xuan Wu District
Beijing 100761
Phone: (86-10) 6320 2036
Fax: (86-10) 6326 6190

Baiguanglu Ertiao No. 1
Guang'an Mennei
P.O. Box 2905
Beijing 100761
Minister: Yang Zhenhuai
Phone: (86-1)365563
Telex: 22455 MWREP CN

Bureau of Hydrology
Baiguanglu Ertiao No. 1
Guang'an Mennei
P.O. Box 2905
Beijing 100761
Director: Hu Zongpei
Phone: (86-1)365331
Telex: 22466 MWREP CN

Nanjing University
Department of Earth Sciences
No. 22, Hankou Road
Nanjing 210093
Chair: Chen Jun
Phone: (86-025) 6637651-2925
Fax: (86-025) 3302728
E-Mail: postgeo@nju.edu.cn

National Bureau of Surveying and Mapping (C)
9 Sanlihe Road, Baiwanzhuang
Beijing 100830
Director General: Jin Xiangwen
Phone: (86-10)6831-1564
Fax: (86-10)6831-1564

National Environmental Protection Agency
Fu Wai Avenue
Beijing 100037
Phone: (86-10) 6833 2299-4100
Fax: (86-10) 6831 4675

National Geomatics Center of China (NGCC)
1 Baishengcun, Zizhuyian, Haidian District
Beijing 100044
Phone: (86-10)6842-4074
Fax: (86-10)6842-4101

Northwest University
Department of Geology
North Taibai Road, No. 1
Xi'an 710069
Chair: Mei Zhi-chao
Phone: (86-029) 7215036-2202
Fax: (86-029) 7212323

Palaeontological Society of China
Nanjing Institute of Geology and
 Palaeontology
Academia Sinica
Chi-Ming-Ssu
Nanjing 210008
Wu Wang-Shi
Fax: 86/25/714182

Peking University
Department of Geophysics, Haidian
Beijing 100871
Chair: Zang Shao-xian
Phone: (86-010) 2501141
Fax: (86-010) 2564095
E-Mail: sxzang@earth.geop.pku.edu.cn

Peking University
Department of Geology
Beijing 100871
Chair: Li Mao-song
Phone: (86-010) 2501150
Fax: (86-010) 2501187
Telex: 22239PKUNI CN

University of Petroleum
Shuiku Road, Changping
Beijing 102200
Chair: Zhang Si-wei
Phone: (86-010) 9745566 EXT. 3074
Fax: (86-010) 9744849
Telex: 6974BEIJING
E-Mail: mx%"sdxyq@bepc2.ihep.ac.cn"

Research Institute of Petroleum (R)
Exploration and Development
20 Xueyuanlu, Haidian District
Beijing 100083
President: Shen Pingping
Phone: (86-10)62097451
Fax: (86-10)62097181
Telex: 22007 CCLBJ CN
E-Mail: riped@public.bta.net.cn

State Oceanic Administration (G)
1 Fuxing Menwai
Beijing 100860
Director General: Yan Hongmo
Phone: (86-10)6853 2211 - 5704
Fax: (86-10)6853 3515
Telex: 22536 NBO CN

State Seismological Bureau (IGSSG) (G)
63 Fuxing Lu
Beijing 100036
Director: Fang Zhangshu
Phone: (86-1)811928
Telex: 222351 SSB CB

Institute of Crustal Dynamics
Xi San Qi
Beijing 100037
Chair: Zhao Guo-guang
Phone: (086-010) 291 3866

Institute of Engineering Mechanics
9 Xuefulu, Harbin
Heilongjiang Province
Director: Xie Junfei
Phone: (86)62901

Institute of Geology
Qijiahuozi, Desheng Menwai
Beijing 100029
Honorary Director: Ma Zongjin
Phone: (86-1)235-6421
E-Mail: disastg@mimi.cnc.ac.cn

Institute of Geophysics
5 Minzuxueyyan Nanlu, Haidian District
Beijing 100081
Director: Chen, Yuntai
Phone: (86-10)68417744
Fax: (86-10)68415372
Telex: 221032IGSSB CN
E-Mail: chenyt@cdsndmc.css.gov

Institute of Seismology
Xiaohongshan, Wuchang
Wuhan, Hebei Province
Director: Zhu Yucheng
Phone: (86)813513/813412
Telex: 87051

Tongji University
Department of Marine Geology and
 Geophysics
1239, Siping Road
Shanghai 200092
Chair: Wang Jia-lin
Phone: (86-21) 5455080
Fax: (86-21) 5458965
E-Mail: wjldmg@tju.ihep.ac.cn

Wuhan Technical University of Surveying and Mapping
39 Luoyu Road, Wuhan
Hubei Province 430070
Phone: (86-27)786-1845
Fax: (86-27)786-5973

Zhingshan University
College of the Earth and Environmental Sciences
West Xinggang, No. 135
Guangzhou 510275
Chair: Luo Hui-bang
Phone: (86-020) 4186300-2495
Fax: (86-020) 4429173
Telex: 8775
E-Mail: leiy@bepc2.ihep.ac.cn

COLOMBIA

Colombian Volcano Observatory
Observatorio Vulcanologico de Colombia
Avenida 12 de Octubre No. 15-47
Apartado Aereo 1296
Manizales
Fernando Gil Cruz
Phone: 68-843004
Fax: 68-826735

Empresa Colombiana de Minas Ecominas (G)
Calle 12 3-07
Bogota, D.E.
Director: Vincente Giordanelli
Phone: (57-1)2877136
Fax: (57-1)2874606

Instituto de Hidrología, Meteorología, y Estudios Ambientales (C,E,G,H,R)
Carrera 5a. No. 15-80, Piso 18
Apartado Aéreo 18633
Santafe de Bogota
Director General: Pablo Leyva
Phone: (57-1)2860658
Fax: (57-1)2860658

Instituto de Investigaciones en Geosciencias, Mineria y Quimica (INGEOMINAS) (C,G,H)
Diagonal 53 No. 34-53
Apartado Aereo 4865
Bogota
Director: Adolfo Alarcón Guzman
Phone: (57-1)2221811
Fax: (57-1)2220797/2223597
Telex: 2220887
E-Mail: a.alarcon@ingeomin.gor.co

Instituto Geografico "Agustin Codazzi" (C)
Carrera 30 NO. 48-51
Santafede Bogota D. C.
Director: Santiago Borrero Mutis

Phone: (57-1)3680960
Fax: (57-1)3681029
E-Mail: nmyard@itecs5.telecom_co.net

Nacional de Metales Preciosos S.A. (R)
Carrera 13 6-45
Bogota, D.E.
Director: Jorge Bendeck
Phone: (57-1)2839891
Fax: (57-1)2811885
Telex: 42648

CONGO

Cadastre du Congo (C)
B.P. 544
Brazzaville

Central African Mineral Resources Development Centre
Centre de Mise en Valeur des Resources Minerales de l'Afrique
B.P. 579
Brazzaville

Institut Geographique (C)
B.P. 125
Brazzaville

Service des Mines et de la Geologie (G)
B.P. 2124
Brazzaville

Service Topographique et du Cadastre (C)
Agence de Brazzaville
B.P. 125 Route du Djove
Brazzaville
Director: Antoine Ondima

COOK ISLANDS

Department of Survey and Physical Planning (C)
Secretary of Survey and Physical Planning
Raratonga
Secretary: G. Cowan

Office of the Prime Minister (G)
Private Bag
Avarua
Rarotonga
Scientific Research Officer: Temu Okotai
Phone: (682)21-150
Fax: (682)23-792

COSTA RICA

Commission on Geographic Information Systems
Universidad Nacional
Escuela de Ciencias Geograficas
Apartado 86
Heredia 3000
M. Lyew

Consejo Nacional de Investigaciones Cientificas y Technologicas (CONICIT) (G)
Apartado 10318
1000 San Jose
President: Rodrigo Zeledon

Universidad de Costa Rica
Escuela de Geologia
San Pedro
San Jose 35-2060
Director: Sergio Paniagua Hernandez
Phone: (506)225 7941
Fax: (506)234 2347

Departamento de Desarrollo Geologico y Recursos Minerales (G)
Corporacion Costarricense de Desarrollo (CODESA)
Apartado 10254
1000 San Jose
Director: Dr. Alfonso Monge

Direccion de Geologia y Minas (G)
Ministerio del Ambiente y Energia
Apartado 10104-100
San Jose
Director: JoseFrancisco Castro Muñoz
Phone: (506)233-23-60
Fax: (506)233-23-34

Instituto Costarricense de Electricidad (ICE) (G)
Departmento de Geologia
Apartado 10032
San Jose, 1000
Chief: Allan Lopez S.
Phone: (506)207141
Fax: (506)314737/314744
Telex: 2140 ICE

Instituto Geografico Nacional (C,G)
Apartado 2272
1000 San Jose
Director: Fernando M. Rudin Rodriguez
Fax: (506)257 52 46

Ministerio de Ciencia y Tecnologia (G)
Apartado 5589
1000 San Jose
Minister: Eduardo Sibaja Arias
Phone: (506)231-5088
Fax: (506)296-3700
E-Mail: micit@sol.racsa.co.cr

Ministerio de Recursos Naturales, Energia y Minas (G)
Apartado 10104
1000 San Jose
Senior Director of Mining: Ursula Rehagg K.
Phone: (506)334533, EXT. 306
Fax: (506)570697
Telex: 2363 ENERGIA CR

Servicio Nacional de Aguas Subterraneas (SENAS) (H)
Apartado 5262
1000 San Jose
Executive Director: A. Suarez

Volcanologic and Seismologic Observatory of Costa Rica
Universidad Nacional
Ovsicori-Una
Apartado 86-3000
Heredia
Eduardo Malavassi
Phone: (506) 37-4570
Fax: (506) 38-0086

COTE D'IVOIRE

Centre de Cartographie et de Télédétection de la DCGTx (DCGTx/CCT) (C)
Direction et Control des Grand Travaux
01 BP 3862
Abidjan 01
Director: Kouadio Konan
Phone: (225) 44 22 04
Fax: (225) 44 28 86

Direction des Hydrocarbures (G)
B.P. V 42
Abidjan
Director: Rene Brancart

Direction de la Geologie
B.P. V 28
Abidjan
Director: Victor Sea

Direction des Mines et de la Geologie (G)
B.P. V 81
Abidjan
Director: Jean Yagui Likan

Ministere des Mines (C,G)
B.P. V 50
Abidjan
Minister: Dr. Yed E. Angoran
Telex: 0983 22262 CI

Service du Drainage et de l'Assainissement (H)
B.P. 21 181
Abidjan

Societe Nationale d'Operations Petroleres de la Cote d'Ivoire (PETROCI) (G)
B.P. V 194
Abidjan
Director-General: Paul Ahui

Societe pour la Realisation de Forages d'Exploitation en Cote d'Ivoire (H)
Ministere du Plan
B.P. V 65
Abidjan
Director: Antoine Kouao

Societe pour le Developpement Minier de la Cote d'Ivoire (SODEMI) (C,G)
B.P. 2816
Abidjan
Director: Joseph N'zi

CUBA

Instituto Cubano de Geodesia y Cartografia (G)
Loma y 39
Nuevo Vedado
La Habana
Director: Emilio Lluis Rojo

Instituto Cubano de Hidrografia (H)
Calle 11, NO. 514, 4 2
Miramar
Director: A.J.C. Hernandez

Instituto de Geofisica y Astronomia (G)
Calle 212, No. 2906
Reparto Cubanacan
La Habana 16
Director: R.A. Morales

Instituto de Geografia (H)
Calle 11, No. 514
La Habana 4
Director: Glaston Oliva

Instituto de Geologia (G)
Academia de Ciencias de Cuba
Ave. Van-Troi no. 17203
Rancho Boyeros
Apartado Postal 10
La Habana
Director: Dr. J.F. de Albear Franquiz

Instituto de Geologia y Paleontologia (G)
Calzada No. 851
La Habana 4
Director: Lenia Montero Zamora

Instituto de Oceanologia (G)
Avenida Ira No. 18406
Cubanacan
La Habana 16
Director: Bienvenido Man'n Lambrana
E-Mail: oceano@ceniai.cu

Instituto Nacional de Recursos Minerales (ICRM) (G)
O'Reilly y Aguacate
La Habana
Director: Salvador Salas

CYPRUS

Department of Lands and Surveys (C)
Ministry of the Interior
Nicosia
Director: Andreas Kotsonis
Phone: (357-2)302210
Fax: (357-2)446056

Environmental Service (E)
Ministry of Agriculture, Natural Resources and Environment
Nicosia, CY411
Director: Nikos Georgiades
Phone: (357-2)302883
Fax: (357-2)363945

Geological Survey Department (E,G)
Levkonos 1
Strovolos
Nicosia
Phone: (357-2)302013
Fax: (357-2)316873

Mines Service (R)
Ministry of Commerce and Industry
Nicosia
Head: Glafkos Kronides
Phone: (357-2)302209
Fax: (357-2)367856

Water Development Department (H)
Ministry of Agriculture and Natural Resources
Nicosia
Director: C. A. Christodoulou
Phone: (357-2)303303
Fax: (357-2)445019

CZECH REPUBLIC

Czech Geological Survey (E,G,H,R)
Klárov 3
118 21 Praha 1
Director: Dr. Zdenek Kukal
Phone: (42-2)24002206
Fax: (42-2)24510480
http://www.cgu.cz

Geodetic and Cartographic Administration (C)
Arbesovo Nam. 4
Praha 5 - Smichov
Director: Ing. Jaromir Karnold
Phone: (42-2)530716
Telex: (42-2)121471

Geodetic Institute (C)
Kostelni 42
170 30 Praha 7 - Holesovice
Director: Miroslav Miksovsky
Phone: (42-2)371441

Geological Institute AS CR (E,G)
Rozvojova135
165 02 Praha 6 - Lysolaje
Director: V. Houša
Phone: (42-2)24311421
Fax: (42-2)24311578
http://www.gli.cas.cz

International Association on the Genesis of Ore Deposits
Czech Geological Survey
Box 65
790 01 Jesenik
Jaroslav Aichler
Phone: (45-2) 3294
Fax: (45-2) 533564

Military Geographical Institute (C)
Rooseveltova 23
150 01 Praha 6
Phone: (42-2)3122712

Vyzkumny Ustav Geodeticky, Topograficky a Kartograficky (C)
250 66 Zdiby 98
Praha - Vychod

Director: Ing. Hynek Wolf
Phone: (42-2)896131

DENMARK

Danish Land Development (H)
Hydrological Survey
Anholtvej 4
DK 4200 Slagelse
Chief Hydrologist: J.L. Jensen
Phone: (45-5)3521701
Fax: (45-5)8101702

Geodetic Institute (C)
Rigsdagsgaarden 7
DK-1218 Copenhagen K
Director: Flemming Wiinblad

Geological Survey of Denmark and Greenland (G)
Thoravej 31
DK-2400 Copenhagen NV
Director: Ole Winther Christensen
Phone: (45-3)1106600
Fax: (45-3)196868
Telex: 19999 DANGEO DK
E-Mail: geus@geus.dk

International Association of Geodesy
University of Copenhagen
Department of Geophysics
Juliana Maries Vej 30
DK-2100 Copenhagen O
C.C. Tscherning
Phone: (45) 35 32 05 82
Fax: (45) 35 36 53 57
E-Mail: cct@gfy.ku.dk

International Council for the Exploration of the Sea
Palaegade 2-4
DK-1261 Copenhagen K
E.D. Anderson
Phone: (45-3) 15 42 25
Fax: (45-3) 93 42 15

DJIBOUTI

Imst. Sup. d'Estudes et Recherches Scientifique et Tech. (G,H)
B.P. 486
Djibouti City
Director: Anis Abdullah

Phone: (253)352795
Fax: (253)354812
Telex: 5811 DJ

Ministere des Travaux Publiques (G)
P.B. 24
Djibouti City
Minister: Omar Kamil Warsama

Service de l'Hydraulique (C,E,G,H)
Ministere de l'Agriculture et de l'Hydraulique
B.P. 453
Djibouti City
Chief: Mohamed Ismael
Phone: (253)35-68-70
Fax: (253)35-58-79

DOMINICA

Ministry of Agriculture, Trade, Industry and Tourism (C,G,H)
Government Headquarters
Roseau
Director of Surveys: L. Cassell
Phone: (809)4482401
Fax: (809)4485200
Telex: 03948613

DOMINICAN REPUBLIC

Departamento de Ciencias Geograficas (C,E,G)
Facultad de Ciencias
Universidad Autonoma de Santo Domingo
Santo Domingo
Director: Elsa Beatriz Culo Tortarolo
Phone: (809) 686-0448
Fax: (809) 533-1106

Direccion General de Mineria (G)
Edificio "El Huacal", Piso 10
Ave. Mexico
Santo Domingo
Director: Gerald M. Ellis
Phone: (809)6877557
Fax: (809)6868327

Instituto Geografico Universitario (C)
Facultad de Ciencias
Universidad Autonoma de Santo Domingo
Santo Domingo
Director: Orlando Adams
Phone: (809)6898310

Instituto Nacional de Recursos Hidraulicos (H)
Apartado Postal 1407
Santo Domingo
Director: Jose C. Farias Cabral
Phone: (809)5323271

ECUADOR

Center of Integrated Surveys of Natural Resources by Remote Sensing (CLIRSEN) (G)
Edificio Instituto Geografico Militar
P.O. Box 8216
Quito
Executive Director: Myr. Ing. Galo Villacis Rueda
Phone: (593-2)545090

Charles Darwin Research Station (E)
Estacion Cientifica Charles Darwin
Casilla 17-01-3891
Quito
Phone: (593-5) 526146
Fax: (593-4) 564636
http://fcdarwin.org.ec/welcome.html

Department of Geology and Mines
Direccion General de Geologia y Minas
Ministerio de Recursos Naturales y Energeticos
Carrion No. 1016 y Paez
Quito

Direccion General de Hidrocarburos (G)
Calle Santa Prisca 223
Quito
Director General: Ing. Rodrigo Ceron Chamorro
Phone: (593-2)523400

Ecuadorian Geological and Geophysical Society
Casilla 371A
Quito
Peter Hall

Instituto Ecuatoriano de Mineria (INEMIN) (G)
Ministerio de Energia y Minas
10 de Agosto 5540 y Villalengua
Quito
Director General: Econ. Horacio Rueda Andrade
Phone: (593-2)240209

Instituto Geofisico (G)
Escuela Politecnica Nacional
P.O. Box 1701-2759
Quito
Technical Director: M. Hall
Phone: (593-2)225655
Fax: (593-2)567847
E-Mail: mhall@instgeof.ecx.ec

Instituto Geográfico Militar (IGM) (C)
Gral. Paz y Miño y Seniergues
Apartado 17-01-2435
Quito
Phone: (593-2)522066/522148
Fax: (593-2)569097
E-Mail: igm1@igm.mil.ec

Instituto Nacional de Meteorologia e Hidrologia (INAMHI) (E,H)
Av. de los Shyris 1570 y
Naciones Unidas
Edificio Alvarez-Andino
Director: Ing. Franco Rios Castillo

Division de Hydrologia
Departamento de Aguas Subterraneas
Calle Iñaquito 700 y Corea
Quito
Director: Gustavo Gómez A.
Phone: (593-2)433-935
Fax: (593-2)433-934

Instituto Oceanografico de la Armada (INOCAR) (C,E,G)
P.O. Box 5940
Guayaquil
Director: JoseOlmedo Morán
Phone: (593-4)481847
Fax: (593-4)485166
E-Mail: inocar@inocar.mil.ec

EGYPT

Academy of Scientific Research and Technology (G,H)
101 Kasr El Aini Street
Cairo
President: Abou El Fotouh Abdel Latif
Phone: (20-2)3546532
Telex: 03069 ASRT UN

Aerial Survey of Egypt (C)
308 El-Haram Street
Giza
Cairo
Director: Mosaad Ibrahim

Phone: (20-2)852950
Telex: 739794 WAZRA UN/93794 WAZRA UNH

Desert Research Institute (G,H)
Mathaf El Matareyya Street
Cairo
Director: Mohmoud Mounir
Phone: (20-2)2435449

Egypt Petroleum Exploration Society
Trapetco
24 Kamel El Shennawi St.
Box 2528
Data Exchange Box 59
Garden City, Cairo
Salah Hafez

The Egyptian Geological Survey and Mining Authority (G)
3 Saleh Salem Road
Abbassyia
Cairo
Chairman: Dr. Gaben Naim
Phone: (20-2)2828013/2855660
Fax: (20-2)4820128
Telex: 22695 UN GEOSU
E-Mail: baha@frcu.eun.eg

Egyptian Petroleum Research Institute
Nasr City
Cairo

Egyptian Remote Sensing Center (C,G)
101 Kasr El Aini Street
Cairo
Chairman: Ahmed S. Ayoub
Phone: (20-2)3540173/3557110
Fax: (20-2)3557110

Hydraulics Research Institute (H)
Delta Barrage 13621
Kanater-al-Khairia
Director: M.B.A. Saad
Phone: (20-2)2188268
Fax: (20-2)2189539
E-Mail: abuzeid@frcu.eun.eg

National Research Institute for Astronomy and Geophysics (G)
Cairo-Helwan
Director: Hanafy Aly Deebs
Phone: (20-2)782683
Fax: (20-2)782683
Telex: 93070

Petroleum Research Institute (G)
Nasr City
Cairo
Director: Bahram Mahmoud

Egypt

Phone: (20-2)609666/603305
Telex: 93069 ASRT UN

Survey of Egypt (C)
1 Abdel Salam Arif Street
Giza
Cairo
Chairman: Ali Abdel Rahman
Phone: (20-2)3484904/3484422
Telex: 739794 WAZRA UN/93794 WAZRA UNH

Water Research Center (H)
22 Galaa Street
Cairo
Director: Mahmoud A. Abou Zeid
Phone: (20-2)760474
Telex: 20275 WRC UN

EL SALVADOR

Center for Geotechnical Research (G)
Avenida Peralta, Final, Contiguo a Talleres El Coro
San Salvador
Director: Ing. Julio Roberto Salazar Mena
Phone: (503)229011

Energy Superintendency (G)
Comision Ejecutiva Hidroelectrica del Rio Lempa (CEL)
Centro de Gobierno
San Salvador
Chief: Raymundo Cisniega
Phone: (503)719855/229046
Telex: ENERGEL SAL 20301

General Directorate of Irrigation and Drainage (C,H)
Canton El Matazano, Soyapango
San Salvador
Director: Alirio Edmundo Mendoza
Phone: (503)770490

Ministerio de Agricultura y Ganaderia (C,E,H,R)
Direccion General de Recursos Naturales Renovables
Cantón El Matazano, Soyapango
Director: Inés Maria Ortiz
Phone: (503) 294-0566
Fax: (503) 294-0575

Ministrey of Public Works (C,G)
La. Av. Sur No. 630
San Salvador
Ministro: Luis Lopez Ceron
Phone: (503)716026

Ministry of Agriculture (C,H)
Alameda Roosevelt 2823
San Salvador
Minister: Oscar Morales Herrera
Phone: (503)2324434/242944

National Geographic Institute (C)
"Ingeniero Pablo Arnoldo Guzman"
Avenida Juan Bertis No. 79
San Salvador
Director: Ing. Santiago Muricio Garcia Contreras
Phone: (503)255060

Weather Forecast and Hydrology Service (C,H)
Renewable Natural Resources Center (CENREN)
Canton El Matazano, Soyapango
San Salvador
Director General: Ing. Leonardo Merlos Ventura
Phone: (503)270484/270622
WMO/CHY Representative: Dr. Gelio Tomas Guzman

ESTONIA

Geological Survey of Estonia
Eesti Geoloogiakeskus
80/82 Kadaka tee
EE0026 Tallinn
Rein Raudsep
Phone: (372 2) 593964
Fax: (372 2) 6579664
E-Mail: egk@estpak.ee

Tartu University
Institute of Geology
Vanemuise 46
Tartu EE2400
Phone: (372) 7 430607
Fax: (372) 7 430643
E-Mail: geol.@ut.ee

ETHIOPIA

Ethiopian Institute of Geological Survey (EIGS) (G)
P.O. Box 2302
Addis Ababa
General Manager: Sheferaw Demissie
Phone: (251-1)159926
Fax: (251-1)517874/711099
Telex: 21042

Ethiopia

Ethiopian Mineral Resources Development Corp. (G,R)
P.O. Box 2543
Addis Ababa
General Manager: Dr. Tebebe Tafesse
Phone: (251-1)610074
Fax: (251-1)611776
Telex: 21463

Geophysical Observatory
University of Addis Ababa
Box 1176
Addis Ababa
Laike M. Asfaw
Phone: (251) 11-72-53

Ministry of Mines and Energy (G,R)
P.O. Box 486
Addis Ababa
Minister: Izaddin Ali
Phone: (251-1)157413
Fax: (251-1)517874
Telex: 21448
Vice Minister of Mines: Shemsudin Adhmed
Phone: (251-1)152891
Fax: (251-1)517874
Telex: 21448

FIJI

Lands and Survey Department (C)
Ministry of Lands, Local Government and Housing
P.O. Box 2222, Government Buildings
Suva
Director: A. Queet
Phone: (679)211516
Fax: (679)304037
Telex: 2167 FOSEC FJ

Mineral Resources Department (C,E,G,R)
Ministry of Energy and Mineral Resources
Private Mail Bag, G.P.O.
Suva
Phone: (679)381611
Fax: (679)370039
E-Mail: brao@mrd.gov.fj

University of the South Pacific
Earth Science Coursework Programme
P.O. Box 1168
Suva
Fiji
Phone: 679-313-900
Fax: 679-301-305

South Pacific Applied Geoscience Commission
Private Mail Bag GPO
Suva
Philipp Muller
Phone: (679) 381377
Fax: (679) 370040
E-Mail: philipp@sopac.org.fj

United Nations Development Programme
Private Mail Bag
Suva
Phone: (679) 312500
Fax: (679) 301718

Water and Sewerage Section (H)
Public Works Department
Ministry of Communications, Work and Transport
Ganilau House
Private Mail Bag, G.P.O.
Suva
Director: G. Green
Phone: (679)315133
Telex: 2104 FJ

FINLAND

Cartographic Department (C)
Opastinsilta 12
00520 Helsinki
Director: Osmo Niemela
Phone: (358-0)1541
Fax: (358-0)147289
Telex: 125 254 MAP SF

Coordinating Committee on the Himalayan Region
Inter-Union Commission on the Lithosphere
Geological Survey of Finland
02150 Espoo
M.V. Knorring
Phone: (358-0) 46931
Fax: (358-0) 462 205

Exchange Centre for Scientific Literature
Mariankatu 5
FIN-00170 Helsinki

Finnish Environment Institute (E,H)
P.O. Box 140
Fin-00251 Helsinki
Research Director: Juha Kämäri

Phone: (358-0)4030 0771
Fax: (358-0)4030 0790
E-Mail: juha.kamari@vyh.fi

Finnish Environment Institute

Monitoring and Assessment Division
Kesakatu 6
FIN-00260 Helsinki
Division Manager: Pertti Seuna
Phone: (358-0)403000
Fax: (358-0)40300391
http://www.vyh.fi/syke/syke.html.

Geodetic Institute (C)

Geodeetinrinne 2
FIN 02430 Masala
Director: Prof. Juhani Kakkuri
Phone: (358-0)295 55307
Fax: (358-0)295 55200
E-Mail: gl@fgi.fi

Geological Society of Finland

Suomen Geologinen Seura
Betonimiehenkuja 4
FIN-02150 Espoo
Reijo Salminen
Phone: (358-0) 46931
Fax: (358-0) 462-205
E-Mail: reijo.salminen@gsf.fi

Geological Survey of Finland (G)

Betonimiehenkuja 4
FIN 2150 Espoo
Director: Dr. Veikko Lappalainen
Phone: (358-0)46931
Fax: (358-0)462205
Telex: 123 185 GEOLO FI
E-Mail: veikko.1appalainen@gsf.fi

Geophysical Society of Finland

Geofysiikan Seura/Geofysiska Sallskapet
Hydrological Office
National Board of Waters and the Environment
Box 436
00101 Helsinki
Sirkka Tattari
Phone: (358-0) 1929 562
Fax: (358-0) 1929 577

University of Helsinki

Department of Geology
P.O. Box 11 (Snellmaninkatu 3)
FIN-00014 University of Helsinki
Phone: +358-0-191123425
Fax: +358-0-19123466
E-Mail: kirsi-marja.ayras@helsinki.fi

Helsinki University of Technology

Department of Surveying
Otakaari 1
FIN-02150 Espoo
Phone: (358-0) 4511
Fax: (358-0) 465007
E-Mail: teuvo.parm@hut.fi

International Peat Society

Kuokkalantie 4
FIN-40420 Jysja
Phone: (358-41) 674 042
Fax: (358-41) 677 405

National Land Survey of Finland (C)

Opastinsilta 12
00520 Helsinki
Director General: Jarmo Ratia
Phone: (358-0)1541
Fax: (358-0)154 5005
E-Mail: jarmo.ratio@nls.fi

Water and Environment Research Office (E,H)

Pohjoinen Rautatiekatu 21 B
00100 Helsinki
Office Chief: Lea Kauppi
Phone: (358-0)40281
Telex: 126 086

FRANCE

Center for Volcanological Research

Centre de Recherches Volcanologiques
5, rue Kessler
63038 Clermont-Ferrand Cedex
Jean-Francois Lenat
Phone: (33) 73 34 67 46
Fax: (33) 73 34 67 44

Centre National de la Recherche Scientifique (CNRS) (G)

15, quai Anatole France
75700 Paris
President: Claude Frejacques
Phone: (33-1)47531515
Telex: 260034

Commission for the Geological Map of the World (C,G)

Commission de la Carte Geologique du Monde
Maison de la Geologie
77 Rue Claude-Bernard
75005 Paris
Philippe Bouysse
Phone: (33-1) 47072284
Fax: (33-1) 43369518
Telex: 206411 (F) CGMW

Directorate of Geological Survey (E,G,H)

Bureau de Recherches Geologiques et Minieres (BRGM)
Avenue de Concyr, Orleans-La Source
(loiret) - BP 6009
45060 Orleans Cedex 2
Director: Laurent Le Bel
Phone: (33)38 64 34 34
Fax: (33)38 64 35 18
Telex: BRGM 780258 F
Advisor, Environment and Risks: Philippe Masure

P.O. Box 1492
Jeddah 21431
Saudi Arabia
Phone: (966-2) 665-1104/1578
Telex: 601674 BRGM SJ

Division of Earth Sciences

Unesco
1, rue Miollis
F-75732 Paris-Cedex 15
F.W. Eder
Phone: (33-1) 45 68 41 15
Fax: (33-1) 43 06 77 76
E-Mail: w.eder@unesco.org

French Clay Group

Groupe Français des Argiles
Universitede Poitiers
URA 721 CNRS "ASA" Generale et Miniere
40 ave. Recteur Pineau
86022 Poitiers Cedex
Alain Decarreau
Phone: (33) 49 45 33 89
Fax: (33) 49 45 40 17
E-Mail: petit@zeus.univ-poitiers.fr

French Palaeontological Association

Association Paleontologique Francaise
Institute de Paleontologie, Museum National d'Histoire Naturelle
8, rue Buffon
75005 Paris
Mireille Gayet
Phone: (33-1) 40 79 30 10
Fax: (33-1) 40 79 34 84

Geological Society of France

Societe Geologique de France
77, rue Claude-Bernard
75005 Paris
F.P. Rangin
Phone: (33-1) 43 31 77 35
Fax: (33-1) 45 35 79 10

Geological Society of the North

Societe Geologique du Nord
Sciences de la Terre
Batiment SN5
Universite de Lille-Flandres-Artois, F-59655 Villeneuve d'Ascq Cedex
Jean-François Deconninck
Phone: (33) 20 43 41 45
E-Mail: jean-francois.deconinck@univ-lille1.fr

Institut Français du Pétrole (IFP)

1 & 4, avenue de Bois-Préau
BP 311
92506 Rueil-Malmaison Cedex
Chairman: Pierre Jacquard
Phone: (33-1)47 52 60 00
Fax: (33-1)47 52 70 00
Telex: 634202 F

Institut Géographique National (C)

136 bis rue de Grenelle
75700 Paris 07SP
Director General: Jacques Fremiot
Phone: (33-1)43 98 80 00
Fax: (33-1)43 98 84 00
Telex: 204989 F

Intergovernmental Oceanographic Commission

UNESCO
7 Place de Fontenoy
75700 Paris
G. Kullenberg
Phone: (33-1)45683983
Fax: (33-1)40569316
http://www.unesco.org/ioc

International Association of Engineering Geology (E,G)

Laboratoire Central des Ponts et Chaussees
58, Boulevard Lefebvre
F-75732 Paris Cedex 15
L. Primel
Phone: (33-1) 40-43-52-43
Fax: (33-1) 40-43-54-98

International Center for Training and Exchanges in the Geosciences

Centre International pour la Formation et les Echanges Geologiques
B.P. 6517
45065 Orleans Cedex 2
Jean-Claude Napias
Phone: (33) 38 64 33 67
Fax: (33) 38 64 34 72

International Geological Correlation Programme (IGCP)

UNESCO
1, rue Miollis
75732 Paris Cedex 15

Vladislav Babuska
Phone: (33-1) 45 68 41 23
Fax: (33-1) 43 06 17 76
E-Mail: v.babuska@unesco.org

International Union of Geodesy and Geophysics

Union Geodesique et Geophysique
  Internationale
CNES-GRGS-BGI
Av. Edouard Belin 18
F-31055 Toulouse Cedex
G. Balmino
Phone: (33) 6133-2989
Fax: (33) 6125-3098

Society for Geology Applied to Mineral Deposits

Centre de Recherche sur les Matieres
  Premieres Minerales et Energetiques
B.P. 23
54501 Vandoeuvre les Nancy
Maurice Pagel
Phone: (33) 83 44 19 00
Fax: (33) 83 44 00 29
http://www.immr.tu-clausthal.de/sga.html

SPOT Image (C,E,G,H,R)

5 rue des Satellites
BP 4359
31030 Toulouse Cedex
Chairman:Jacques Mouysset
Phone: (33-62)05 62 19 40 40
Fax: (33-62)05 62 19 40 11
Telex: 532079 SPOTIM
http://www.spotimage.fr/

World Organization of Volcano Observatories

Observatoires Volcanologiques
IPGP
B 89, 4 Place Jussieu
75252 Paris Cedex 05
Phone: (33-1) 44 27 24 00
Fax: (33-1) 44 27 24 01

FRENCH GUIANA

Bureau of Geological and Mineral Research (G)

B.P. 42
Cayenne, F.W.I.

GABON

Bureau de Recherches Geologiques et Minieres (BRGM) (C,G)

B.P. 175
Libreville
Director: Etienne Wilhelem
Phone: (241)760609/764498
Telex: 5576 GO

Direction Generale de Mines et de la Geologie (C,G,H,R)

Ministere des Mines, de l'Energie et du
  Petrole
B.P. 576
Libreville
Director: Ngakoussou Appana Ferdinand
Phone: (241) 76.35.56
Fax: (241) 72.49.90
Telex: 5352 GO/5499 GO

GAMBIA

Department of Lands and Surveys (C)

Cotton Street
Banjul
Director: B.B. Barry Mactarr O. Jobe
Phone: (220)27337

Department of Water Resources (H,R)

7 Marina Parade
Banjul
Director: P.A. Cham
Phone: (220)228216
Fax: (220)225009

Ministry of Agriculture and Natural Resources (E,H,R)

The Quadrangle
Banjul
Permanent Secretary: Sambou Kinteh
Phone: (220)228230
Fax: (220)228998

GERMANY

Alfred-Wegener-Stiftung

Weyerstrasse 34-40
D-50676 Köln
Rolf Meissner
Phone: (49-221)921541-90
Fax: (49-221)9218254
E-Mail: pge65@rz.uni-kiel.d400.de

European Geophysical Society

EGS Office
Max-Planck-Str. 1
D-37911 Katlenburg-Lindau
A.K. Richter
Phone: (49) 5556-1440
Fax: (49) 5556-4709
E-Mail: egs@lina+2.mpae.gwdg.de

Federal Institute for Geosciences and Natural Resources (Bundesanstalt fur Geowissenschaften und Rohstoffe [BGR]) (C,E,G,H,R)

P.O. Box 510153
D-30631 Hannover
President: Martin Kuersten
Phone: (49-511)643-0
Fax: (49-511)643-3
Telex: 9-23730 BGRHAD
E-Mail: 6010spannbru@rzvax.hannover.bgr.de

GDMB Society for Mining, Metallurgy, Resource and Environmental Technology (E,G,R)

Box 10 54
D-38668 Clausthal-Zellerfeld
D. Dornbusch
Phone: (49) 05323/93790
Fax: (49) 05323/937937
http://www.tu-clausthal.de/gdmb

Geological Association (G)

Geologische Vereinigung e.V.
Geschaeftsstelle
Vulkanstr. 23, D-56743
Mendig
Treasurer: Carl D. Cornelius
Phone: (49) 2652/1508
Fax: (49) 2652/52537

Geological Survey of Lower Saxony

Alfred-Bentz-Haus, Postfach 51 01 53
3000 Hannover 51
Fax: (49-511) 6 43-23 04

German Geological Society

Deutsche Geologische Gesellschaft
Stilleweg 2
D-30655 Hannover
President: Hubert Miller
Phone: (49-511) 643-2507
Fax: (49-511) 643-2304

German Geophysical Society

Deutsche Geophysikalische Gesellschaft
Institut für Geophysik und Meteorologie
Albertus-Magnus-Platz
50923 Köln
Fritz M. Neubauer

Phone: (49) 221 470 2310
Fax: (49) 221 470 5198
http://sdac.hannover.bgr.de/dgg/dgg.html

German Mineralogical Association

Deutsche Mineralogische Gesellschaft
Mineralogisch-Petrologisches Institut
Poppelsdoefer Schloss
D-53115 Bonn
Stephan Hoernes
Phone: (49-228) 73 2733
Fax: (49-228) 73 2763
E-Mail: stephan.hoernes@uni-bonn.de

Institute for Mineralogy

Institut fur Mineralogie
Universität Münster
Corrensstr. 24
D-48149 Münster
Phone: (49-251) 83 34 51
http://www.uni-muenster.de/chemie/mi/

International Mineralogical Association

University of Marburg
Institute of Mineralogy
D-35032 Marburg
Stefan S. Hafner
Fax: (49) 06421-288919
E-Mail: hafner@mailer.uni-marburg.de

Paleontological Society

Palaontologische Gesellschaft
Forschungsinstitut Senckenberg
Senckenberganlage 25
D-60325 Frankfurt am Main
Thomas Jellinek
Phone: (49-69) 97075-139
Fax: (49-69) 97075-137
E-Mail: tjelline@sngkw.uni-frankfurt.de

GHANA

Geological Survey of Ghana (G)

P.O. Box M 80
Accra
Director: G.O. Kesse
Phone: (233-21)226490

Hydrological Branch (H)

Public Works Department
P.O. Box 136
Accra
Chief Hydrological Engineer: S.A. Acheampong
Phone: (233-21)6658665

Ghana

Industrial Research Institute (G)
Council for Scientific and Industrial Research (CSIR)
P.O. Box M 32
Accra
Acting Director: R.B. Lartey
Phone: (233-21)775202/776991
E-Mail: iri@ghastinet.gn.apc.org

Water Resources Institute (E,H)
P.O. Box M-32
Accra
Acting Director: A.T. Amuzu
Phone: (233-21)775351
Fax: (233-21)777170
E-Mail: wrri@ghastinet.gn.apc.org

West African Science Association
Box 7
Legon

GREECE

Division of Land Improvement Planning and Exploration and Land/Water Resources (H)
46 Halkokondyli Street
GR-104 38 Athens
Director: Christos Papavassiliou
Phone: (30-1)5243339
Telex: 221734 YDAG

Geological Society of Greece
50 Dodekanissou St.
16231 Byron
Athens

Hellenic Army Geographic Service (C)
3 Evelpidon Street
GR-113 62 Athens
Director: General Peter Kletsas
Phone: (30-1)8842811
Telex: 226379

Institute of Geology and Mineral Exploration (IGME) (C,E,G,H,R)
70 Messoghiou Avenue
Athens 11527
Director: G.C. Katsamangos
Phone: (30-1)7771438
Fax: (30-1)7752211
Telex: 216357 IGME GR
http://lapis.igme.ariadne-t.gr

Land Improvement Works Division (D7) (H)
Ministry of Environment, House Planning

9 Fanarioton Street
GR-101 78
Athens
Director: Ioannis Leontaritis
Phone: (30-1)6445132
Telex: 215018 YDER

Public Petroleum Corporation (DEP) (G)
54 Acadimias Street
GR-106 79 Athens
Director: Constaninos Papaspyridis
Phone: (30-1)6501340-9
Telex: 215255 EMY

GREENLAND

Geological Survey of Greenland (G)
Ostervoldgade 10, Tr. KL
DK-1350 Copenhagen K
Denmark
Director: Martin Ghisler
Phone: (45-1)33118866
Fax: (45-1)33935352
Telex: 19066 GGUTEL

GUADELOUPE

Arrondissement Mineralogique de la Guyane (G,H)
B.P. 448
Pointe-a-Pitre

Guadeloupe Volcano Observatory
Observatoire Volcanologique Guadeloupe
Observatoire Volcanologique de la Soufriere
Le Houëlmont
97113 Gourbeyre
West Indies
Michel Feuillard
Phone: (590) 99-11-33
Fax: (590) 99-11-34

GUATEMALA

Environmental National Committee (E)
7a. Avenida 4-35, Zona 1
Ciudad de Guatemala
Coordination: Ing. Mirella Archila Serrano
Phone: (502-2)21816/532477/535108

43

Guatemala

Guatemalan Geological Society
Sociedad Geologica de Guatemala
Apartado 38
Mixco
Miguel Carballo

Institute of Seismology, Volcanology, Meteorology and Hydrology (INSIVUMEH) (H)
7a Avenida l4-57, Zona l3
Ciudad de Guatemala
Director: Sr. Eddy Sanchez B.
Phone: (502-2)324722/314986
Fax: (502-2)315005
Telex: 363944
E-Mail: mariovsungar.sun.com

Military Geographic Institute (C)
Avenida de las Americas 5-76, Zona l3
Director: Coronel Joaquin Humberto Maldonado de Los Angeles
Fax: (502-2)313548

Geology Division
Avenida Las Americas 5-76, Zona l3
Director: Ing. Edgar Lam Echeverria
Fax: (502-2)313548

Ministry of Energy and Mines (R)
Diagonal 17, 29-78, Zona 11
Ciudad de Guatemala
Minister: Ing. Cesar Augusto Fernandez
Phone: (502-2)760682

GUINEA

Ministry of Agriculture (H)
B.P. 576
Conakry
Director General of Waters and Forests: Khalidou Diallo
Phone: (224-4)461740

Ministry of Mines and Geology (C,G)
B.P. 295
Conakry
Director National: Ibrahima Sory N'Diaye
Phone: (224-4)46-20-11
Fax: (224-4)41-49-13

GUINEA-BISSAU

Ministry of Natural Resources (G,H)
Caixa Postal 399
Bissau
Minister: Filinto Barros
Phone: (245)212618
Director General: Carlos Dias

GUYANA

Guyana Geology and Mines Commission (G)
Upper Brickdam
P.O. Box 1028
Georgetown
Commissioner: Brian Sucre
Phone: (592-2)253047
Fax: (592-2)253047
Telex: GUY GEOL 3042

Hydrometeorological Station (H)
Ministry of Agriculture
18 Brickdam
Stabroek, Georgetown
Chief Hydrometeorological Officer: Sheik Khan
Phone: (592-2)54207
Fax: (592-2)61460

Institute of Applied Science and Technology (G)
University Campus
Turkeyen
East Coast Demerara
Guyana
Director: Dr. Ulric Trotz
Phone: (592-2)53921/650723
Fax: (592-2)53042

Lands and Surveys Department (G)
22 Upper Hadfield Street
D'urban Backlands
Greater Georgetown
Commissioner of Lands and Surveys: Abhai Kumar Datadin
Phone: (592-2)72582/60524-9
Fax: (592-2)64052

Natural Resources Agency (G)
41 Brickday & Boyle Place
Georgetown
Executive Chairman: Winston King
Phone: (592-2)66549/56720/56111
Telex: 3010 GRNA GY

HAITI

Department of Mines (G)
Briand Lafalaise

Delmas 19
P.O. Box 2174
Port-au-Prince
Director General: Pierre Mathurin
Phone: (509-1)60324

Service de Geodesie et de Cartographie (C)

Cite de l'Exposition
Boulevard Harry Truman
Port-au-Prince
Director: Raymond B. Oriol
Phone: (509-1)23225

Service Meteorologique et Hydrologique (H)

Ministere de l'Agriculture, des Resources et du Developpement Rural
P.O. Box 1441
Port-au-Prince
Director: Ing. Waldeck Demetrius
Phone: (509-1)24057

HONDURAS

Direccion General de Minas e Hidrocarburos (C,G)

Apartado Postal 981
Tegucigalpa, D.C.
Director General: Rigoberto Rodriguez
Phone: (504)327848
Fax: (504)325375
Telex: 1404 SERENA HO

Instituto Geografico Nacional (IGN) (C)

Edificio SECOPT
Barrio La Bolsa
Comayaguela, D.C.
Director General: Mayor y Lic. Jose Fausto Aguero
Phone: (504)337403

Oficina Nacional del Catastro (C)

Edificio Didemo, 4 Piso
Tegucigalpa, D.C.
Director: Rodolfo Sthesman
Phone: (504)332081

Servicio Autonomo Nacional de Acueductos y Alcantarillados (SANAA) (H)

Apartado Postal 437
Tegucigalpa, D.C
Director General: Ing. Luis Moncada Gross
Phone: (504)228550

Servicios Hidrologicos y Climatologicos (H)

Ministerio de Recursos Naturales
Apartado Postal 1839
Tegucigalpa, D.C.
Chief: Ing. Jose C. Figueroa Urteche

HONG KONG

Department of Geography and Geology (E,G,H)

University of Hong Kong
Kowloon
Phone: (852)2859 2835
Fax: (852)2559 8994
http://www.hku.hk/geo

Geological Society of Hong Kong

c/o Department of Geography and Geology
University of Hong Kong
Pokfulam Road
R. B. Owen
Phone: (852) 339-7188
Fax: (852) 338-6005

Geological Survey of Hong Kong (G)

Geotechnical Control Office
Empire Centre, Sixth Floor
68 Mody Road, Tsim Sha Tsui East
Kowloon

Geotechnical Engineering Office

Civil Engineering Department
15/F Civil Engineering Building
101 Princess Margaret Road
Homantin
Kowloon
Phone: (852)27625087
Fax: (852)27150501

Hong Kong Polytechnic University

Department of Land Surveying and Geo-Informatics
Hong Kong, Kowloon
Phone: (852) 2766-5968
Fax: (852) 2330-2994
http://www.ls.polyu.edu.hk/

Mines Division (G)

Canton Road Government Offices
Canton Road
Kowloon
Superintendent of Mines: Borhylen Wang

Royal Observatory (H)

134A, Nathan Road
Kowloon

Phone: (852)2926-8200
Fax: (852)2311-9448
http://www.info.gov.hk/ro/index.htm

Survey Division (C)
Survey and Mapping Office
Murray Building, 14th Floor
Garden Road
Kowloon
Phone: (852) 2848-2723
Fax: (852) 2521-8726

HUNGARY

Carpathian-Balkan Geological Association Commission on Stratigraphy, Paleogeography and Paleontology (G)
Geza Csaszar
Pf. 106
H-1442 Budapest
Phone: (36-1) 2510889
Fax: (36-1) 2510703
E-Mail: csaszar@vax.mafi.hu

Central Geological Bureau of Hungary
Kozponti Foldtani Hivatal
Box 374
H-1371 Budapest
Gyorgy Komlossy
Phone: (36-1) 112-7465
Fax: (36-1) 112-2870

Eötvös Loránd Geophysical Institute of Hungary (G)
H-1145 Columbus u. 17/23
Budapest
Director: Tamás Bodoky
Phone: (36-1)252-4999
Fax: (36-1)163-7256
E-Mail: h6124bod@ella.hu

Eötvös Loránd University
  Department of Applied and
    Environmental Geology
  Muzeum KRT y/a
  Budapest 1088
  Phone: (36-1) 266-4992
  Fax: (36-1) 266-4992

  Department of Geology
  Mùzeum KRT y/a
  Budapest 1088
  Phone: (36-1) 266-3956

  Department of Petrology and
    Geochemistry
  Muzeum KRT y/a
  Budapest 1088

Phone: (36-1) 266-3932

Geodesy and Cartography Bureau (C)
H-1442 Bosnyak ter 5
Budapest
Chief: Gyula Biro
Phone: (36-1)635260
Telex: 635260

Geophysical Exploration Co. (G)
Városligeti fasor 42
1068 Budapest
Managing Director: András Zelei
Phone: (36-1)351 1840
Fax: (36-1)351-1840
Telex: 22 55 86
E-Mail: zelei@ges.hu

Hungarian Geological Institute
Nepstadion u. 14, Pf. 106
1442 Budapest

Hungarian Geological Society
Magyarhoni Foldtani Tarsulat
Fo u. 68. 1.102
H-1027 Budapest
Tibor Kecskemeti
Phone: (36-1) 201-9129
Fax: (36-1) 201-9129

Hungarian Geological Survey (G,H)
H-l442 Bp Stefania ut 14
Box 106
Budapest
Director: Karoly Brezsnyanszky
Phone: (36-1)251-4680
Fax: (36-1)251-0703
E-Mail: geo@vax.mafi.hu

Hungarian Hydrocarbon Institute (G)
H-2443 Szazhalombatta
Pf: 32
Budapest
Director: Dr. Sandor Doleschall
Phone: (36-1)800122
Telex: 226636

Hungarian Hydrological Society
Box 433
1371 Budapest

Ministry of Environmental Protection and Water Management (H)
H-1094 Bp. F0
Budapest
Minister: Laszlo Marothy
Phone: (36-1)154840
Telex: 224879

National Crude Oil and Natural Gas Trust (G)
H-1117 Schonherz Z. u. 18
Budapest
Chairman: Dr. Istvan Zsengeller
Phone: (36-1)664000
Telex: 224762

Oil and Gas Laboratories and Engineering Group (G)
P.O. Box 43
H-1311 Budapest
István Bérczi
Phone: (36-1) 180-0122
Fax: (36-1) 180-2369
E-Mail: iberczi@mail.datanet.hu

Seismological Observatory of HAS (G)
H-1112 Meredek 18
Budapest
Head of Department: Gyozo Szeidovitz
Phone: (36-1)319-3382
Fax: (36-1)319-3385
http://www.seismology.hu

Water Resources Research Centre (VITUKI) (H)
H-1095 Kvassay Jeno u. 1
Budapest
Director General: Ödön Starosolszky
Phone: (36-1)215-6140
Telex: 224959
E-Mail: starosolszky@atlmail..com

ICELAND

Geophysics Division (G)
Science Institute
University of Iceland
Dunhagi 3
107 Reykjavik
Director: Leo Kristjansson
Phone: (354)5254800
Fax: (354)5528801
http://www.raunvis.hi.is/~leo

Geoscience Society of Iceland
National Energy Authority
Grensasvegi 9
IS-108 Reykjavik

Iceland Geodetic Survey (C)
Laugavegi 178
105 Reykjavik
Director: Agust Guomundsson

Phone: (354)533 4000
Fax: (354)533 4011
E-Mail: office@lmi.is

Iceland Glaciological Society
Box 5128
IS-125 Reykjavik

Iceland National Energy Authority (G)
Grensasvegur 9
108 Reykjavik
Director General: Jakob Bjornsson
Phone: (354-1)696000
Fax: (354-1)688896
Telex: 2239 ORKUST IS

Geothermal Division
Grensasvegur 9
108 Reykjavik
Director: Gudmundur Palmason
Phone: (354)569-6000
Fax: (354)568-8896
http://www.os.is

Icelandic Institute of Natural History (C,E,G)
Hlemmur 3-5
P.O. Box 5320
125 Reykjavik
Director: Jon Gunnar Ottosson
Phone: (354)562-9822
Fax: (354)562-0815
http://www.nattfs.is/ni-home.htm

Marine Research Institue (G)
Hafrannsóknastofnunin
Skúlagata 4, P.O. Box 1390
121 Reykjavik
Director: Jakob Jakobsson
Phone: (354)552-0240
Fax: (354)562-3790
http://www.hafro.is/hafro/sjalf.html

Nordic Volcanological Institute (G)
University of Iceland
Grensásvegur 50
IS-108 Reykjavik
Director: Gudmundur E. Sigvaldason
Phone: (354-)5254491
Fax: (354-)5629767
http://www.norvol.hi.is/index.html

Surtsey Research Society
Surseyjarfelagid
Box 352
IS-121 Reykjavik
Sveinn P. Jakobsson
Fax: (354)562 9822
E-Mail: sjak@nattfs.is

INDIA

Central Fuel Research Institute (G)
Council of Scientific and Industrial Research
P.O. FRI., District Dhanbad-828108
Bihar
Director: Prof. R. Haque
Phone: (91)61807
Telex: 0629-201 CFRI IN

Central Ground Water Board (H)
Ministry of Water Resources
Jam Nagar House, Man Singh Road
New Delhi 110001
Chairman: D.K. Dutt
Phone: (91-11)382256
Telex: 62941 ND

Central Mining Research Station (G)
Barwa Road
Dhanbad-826001
Director: Dr. B. Singh
Phone: (91-326)2616
Telex: 0629-208 CMRS IN

Central Water Commission (H)
Dewa Bhavan, R.K. Puram-66
New Delhi 110011
Chairman: M.A. Chitale
Phone: (91-11)608312
Telex: 31-66323 CWC IN

The Centre for Earth Science Studies (C,E,G,R)
Akkulam, Thiruvikkal P.O.
Trivandrum, 695031
Director: K. M. Nair
Phone: (91-471)442187
Fax: (91-471)442280

Department of Science and Technology (G)
Earth Sciences Division
Technology Bhawan of
New Mehrauli Road
New Delhi
Advisor: Dr. Harsh K. Gupta
Phone: (91-11)666076
Fax: (91-11)652731
Telex: 66096 DST IN

Geodetic and Geophysical Surveys (C)
Survey of India
Dehra Dun 248001
Director: V.K. Nagar
Phone: (91-135)24528
Telex: 0585210-G & RB IN

Geological, Mining and Metallurgical Society of India
Department of Geology, University of Calcutta
35 Ballygunge Circular Road
Calcutta 700 019
Ajit Kumar Banerji
Phone: (91) 75-3681, EXT 1

Geological Society of India
P.B. 1922
Gavipuram P.O.
Bangalore 560 019
B.P. Radhakrishna
Phone: (91)6613352
Fax: (91)806613352

Geological Survey of India (G)
27 Jawaharlal Nehru Road
Calcutta 700016
Director General: S.K. Acharya
Phone: (91-33)249-6972/6976
Fax: (91-33)296956

Indian Association of Geohydrologists
Geological Survey of India
4 Chowringhee Lane
Calcutta 700 016

Indian Bureau of Mines (R)
Headquarters Building
Civil Lines
Nagpur 440 001

Indian Geologists' Association
Department of Geology
Panjab University
Chandigarh 160014
Naresh Kochhar
Phone: (91-172) 541-740
Fax: (91-172) 54549
Telex: 395-464 RSIC IN

Indian Institute of Remote Sensing (C,E,G,H)
4, Kalidas Road
P.B. No. 135
Dehra Dun 248 001
Uttar Pradesh
Dean: Prof. S.K. Khan
Phone: (91-135)654583
Fax: (91-135)651987
Telex: 0585-224 NRSA IN

Indian Petroleum Publishers
100/9 Naishvilla Road
Dehra Dun 248001

Institute of Petroleum Exploration
Keshava Deva Malaviya

9-Kaulagarh Road
Dehra Dun 248195
Regional Director: Shri Kuldeep Chandra
Phone: (91-135)623193
Fax: (91-135)625265
Telex: 0585/273 MIPE IN
E-Mail: rdiongc.isnet@axcess.net.in

International Association of Seismology and Physics of the Earth's Interior

National Geophysical Research Institute
Hyderabad 500 007
Harsh K. Gupta
Phone: (91-40)671124
Fax: (91-40)671564
Telex: 0425-7018 NGRI IN
E-Mail: director@ngri.uunet.in

National Geophysical Research Institute (G)

Uppal Road
Hyderabad 560007
Director: Harsh K. Gupta
Phone: (91-40)671124
Fax: (91-40)671564
Telex: 0425-7018 NGRI IN
E-Mail: postmast@csngri.ren.nic.in

National Institute of Oceanography (G)

P.O. Dona Paula
Goa 403004
Director: Erlich Desa
Phone: 91(0)832-221322
Fax: 91(0)832-223340
Telex: 0194-316 MGG IN
E-Mail: ocean@bcgoa.ernet.in

National Remote Sensing Agency (C,G,H)

Department of Space
Balanagar
Hyderabad 500037
Andhra Pradesh
Director: K.J. Hebbar
Phone: (91-842)2625727
Telex: 0155-522

Oil and Natural Gas Commission (G)

Institute of Petroleum Exploration
Tel Bhavan
Dehra Dun 248195
Chairman: Col. S.P. Wahi
Phone: (91-135)24203
Telex: 05951206

Palynological Society of India

Environmental Resources Research Centre
P.B. 1230
Peroorkada P.O., Trivandrum
Kerala PIN-695003
P.K.K. Nair
Phone: (91) 61159

Survey of India (C)

Surveyor General Office
Hathi Bukla Estate
P.O. Box 37
Dehra Dun 248001
Surveyor General of India: Maj. Gen. S. M. Chadha
Phone: (91-135)23468/23467
Telex: 0585218 SURVEYS IN

Wadra Institute of Himalayan Geology (G)

Department of Science and Technology
33 General Mahader Singh Road
Dehra Dun 248001
Director: V.C. Thahur
Phone: (91-135)23052
Fax: (91-135)25212
Telex: 0585 326

Working Group on Flood Basalt Volcanism

Department of Earth Sciences
Indian Institute of Technology
Powai
Bombay 400076
K.V. Subbarao
Phone: (91) 5782545, EXT. 3217
Fax: (91) 5783480

INDONESIA

Agency for the Assessment and Application of Technology (BPPT) (G)

Jl. M.H. Thamrin 8
Jakarta Pusat 10340
Deputy Chairman for Natural Resources: Prof. M.T. Zen
Phone: (62-21)324255
Fax: (62-21)324990
Telex: 61321 BPPT IA

Directorate General of Geology and Mineral Resources (DGGMR) (C,E,G,H)

Department of Mines and Energy (DME)
Prof. Supomo 10
Jakarta 12870
Director General: Adjat Sudradjat

Phone: (62-21)8280773
Fax: (62-21)8297642
Telex: 48603 GSMIA

Directorate of Environmental Geology (DEG)
Jalan Diponegoro No. 57
Bandung 40122
Director: P.H. Silitonga
Phone: (62-22)72603
Fax: (62-22)74767
Telex: 28760 DGT IA

Geological Research and Development Centre (GRDC)
Jalan Diponegoro No. 57
Bandung 40122
Director: Irwan Bahar
Phone: (62-22)703205
Fax: (62-22)702669

Marine Geological Institute (MGI)
Jalan Dr. Junjunan 236
P.O. Box 1301
Bandung 40174
Director: Ismail Usna
Phone: (62-22)632151
Fax: (62-22)617887
Telex: 28404 GEOL BD

Volcanological Survey of Indonesia (VSI)
Jalan Diponegoro No. 57
Bandung 40122
Director: Wimpy S. Tjetjep
Phone: (62-22)72606
Fax: (62-22)702761
Telex: 28816 VULKAN IA
E-Mail: vsimvo@ibm.net

Directorate General of Mines (C,E,R)
Jl Gatot Soebroto Kav 49
Jakarta Selatan
Director General: Kuntoro Mangkusubroto
Phone: (62-21)5250447
Fax: (62-21)5251380
http://www.djpu.com

Directorate of Coal
Director: Moh Busono
Phone: (62-21)82428
Fax: (62-22)614168
Telex: 28279 PPTM BD-IA

Mineral Technology Development Centre
Jalan Jenderal Sudirman 623
Bandung
Director: U.W. Soelistijo
Phone: (62-22)613487

Directorate of Mineral Resources (DMR) (G,R)
Jalan Soekarno-Hatta No. 444
Bandung 40254
Director: Kingking A. Margawidjaja
Phone: (62-22)502721
Fax: (62-22)505809
Telex: 28619 DSMBD IA

Indonesian Coal and Mining Association
Jl. Prof. Dr. Supomo SH No. 10
Jakarta 12870
Joe Widartoyo
Phone: (62-21) 830-3632/828-0763
Fax: (62-21) 830-3632

Indonesian National Aeronautics and Space Institute (C,R)
Remote Sensing Technology Center
Jl. Lapan No. 70. Pekayon
Ps Rebo. Jakarta Timur 13710
Deputy Chairman for Remote Sensing: Mahdi Kartasasmita
Phone: (62-21)87101786
Fax: (62-21)8717715
Telex: 49175 IA

Marine Geological Institute (G)
Jalan Dr. Junjunan 236
Bandung 40174
Aswan Yasin
Phone: (62-22) 632151
Fax: (62-22) 617887
E-Mail: doelmgi@ibm.net

Meteorological and Geophysical Agency
Badan Meteorologi dan Geofisika
Jl. Arif Rahman Hakim 3
Jakarta 10340
S. Soetardjo
Phone: (62-21) 390 9409
Fax: (62-21) 310 7788

IRAN

Geological Survey of Iran (G)
P.O. Box 13185-1494
Tehran
Director: A. Heravi
Phone: (98-21)9171
Fax: (98-21)6009338
Telex: 21510 GSOI IR

Ministry of Water and Power (H)
Tehran
Minister: B.N. Zauganeh
Phone: (98-21)8918
Telex: (98-21)891850

National Cartographic Center (NCC) (C)
Azadi Square
P.O. Box 13185-1684
Tehran
Director: A. Shafaat
Phone: (98-21)6001095
Fax: (98-21)6001971
E-Mail: ncc@dci-iran.com

National Geographic Organization (C)
Tehran
Director: Dr. Malmarian
Phone: (98-21)840111
Fax: (98-21)849960

National Iranian Oil Company (G)
Taleghani Ave.
Tehran
Director: G. Aghazadea
Phone: (98-21)6151
Fax: (98-21)6152092

IRAQ

Arab Geologist Association
Box 1247
Baghdad 12112
Wissam S. Al-Hashimi
Phone: (964-1) 8840006
Fax: (964-1) 7193532

Association of Arab Geologists
National Oil Company
Al-Khudani Square
Baghdad
S. Sherif
Phone: (92-51) 240423
Fax: (92-51) 240223

Iraqi Geologists' Union
Box 6244
1-Mansour, Baghdad
Farazdaq Al-Haddad
Phone: (964-1) 5433788
Fax: (964-1) 8763832

Petroleum and Gas Engineering Department (G)
Petroleum Research Center
Scienific Research Council
P.O. Box 10039 - Jadiriyah
Baghdad
Phone: (964-1)7767261

Seismic Exploration Company (C,G)
Ministry of Oil
P.O. Box 476
Khullani Square
Baghdad
Director: Dr. Hisham Fadhil Abul-Hussain
Phone: (964-1)8871115
Telex: 212208/212204 INCO IK

Seismology Unit (G)
Building Research Center
Scientific Research Council
P.O. Box 255 - Jadiriyah
Baghdad
Phone: (964-1)7768927

State Commission for Dams (C,H)
P.O. Box 5982
Baghdad
Director: Nouri Mahmoud Hasan
Phone: (964-1)4169141
Telex: 211207 DAMS IK

State Establishment for Geological Survey and Mining (C,G,R)
Ministry of Industry and Minerals
P.O. Box 986, Alwiya
Baghdad
Director: Harith K. Mahmoud
Phone: (964-1)7195123
Fax: (964-1)7185450
Telex: 2828 IK

State Establishment for Surveying (C)
Ministry of Agriculture and Irrigation
P.O. Box 5813
Gailani Square
Baghdad
Director: Luai Ahmad Mohammad
Phone: (964-1)8886101
Telex: 212916 SURVEY IK

Water Resources Department (H)
Agricultural and Water Resources Research Center
Scientific Research Council
P.O. Box 2416 - Jadiriyah
Baghdad
Phone: (964-1)7512318

IRELAND

Athlone Regional Technical College
Dept of Mineral Engineering
School of Engineering
Dublin Road
Athlone
Co. Westmeath
Phone: (353) 902-72647
Fax: (353) 902-74529

Ireland

Dublin Institute for Advanced Studies
School of Cosmic Physics
5 Merrion Square
Dublin 2
Phone: (353) 1-6774321
Fax: (353) 1-682003

Geological Survey of Ireland (E,G,H)
Beggars Bush, Haddington Road
Dublin 4
Phone: (353-1)6707444
Fax: (353-1)6681782
Phone: (354-1)793377
Fax: (354-1)789527

Hydrometric Service (H)
Office of Public Works
51 St. Stephen's Green
Dublin 2
Director: Tom Bolger
Phone: (354-1)6613111
Fax: (354-1)6761714

Irish Association for Economic Geology
Geological Survey of Ireland
Beggars Bush
Haddington Road
Dublin 4
Phone: (353-1) 660 9511
Fax: (353-1) 668 1782

Irish Geological Association
c/o Trinity College Dublin
Dublin 2
Phone: (353-1) 6081074
Fax: (353-1) 6711199
E-Mail: wysjcknp@ted.ie

Ordinance Survey (C)
Phoenix Park
Dublin 8
Director of Operations: M.C. Walsh
Phone: (354-1)8206100
Fax: (354-1)8204156

Quaternary Research Association
Department of Geography
Museum Building
Trinity College
Dublin 2
Peter Coxon
Phone: (353-1)6081213
Fax: (353-1)6713397
http://www2.tcd.ie/~pcoxon/qra.html

Queens University
School of Geosciences
University Road
Belfast BT7 1NN

Phone: (0044) 232-245133
Fax: (0044) 232-321280

Trinity College
Department of Geology
Dublin 2
Phone: (353) 1-6772941
Fax: (353) 1-6772694
Telex: 93782

University College, Cork
Department of Geology
Cork
Phone: (353) 21-276871
Fax: (353) 21-275948

University College, Dublin
Department of Geology
Belfield
Dublin 4
Phone: (353) 1-7062331
Fax: (353) 1-2837733
Telex: 32693 UCD E1

University College Galway
Applied Geophysics Unit
Galway
Phone: (353) 91-24411
Fax: (353) 25700
Telex: 50023 UCG EL

Department of Geology
Galway
Phone: (353) 91-24411
Fax: (353) 91-25700
Telex: 50023 UCG EL

ISRAEL

Ben-Gurion University of the Negev
Jacob Blaustein Institute for Desert Research
Sede Boker Campus - 84990
Phone: 07-461289
Fax: 07-235646

Earth-Sciences Research Administration (E,G,H)
Ministry of Energy and Infrastructure
234 Jaffa Road
Jerusalem 94383
Director: Dr. Michael Beyth
Phone: (972-2)316129
Fax: (972-2)381444
E-Mail: beyth@netvision.net.il

Geological Survey of Israel (E,G,H)
30 Malkhe Israel Street
Jerusalem 95501
Director: Gideon Steinitz

Phone: (972-2)314211
Fax: (972-2)380688

Hebrew University of Jerusalem
The Institute of Earth Sciences
Givat Ram
Jerusalem
Phone: (972-2) 6584686
Fax: (972-2) 662581

Hydrological Service (H)
Ministry of Agriculture, Water Commission
Yirmiyahu Street, 50
Jerusalem 91060
Director General: S. Kesler
Phone: (972-2)381101
Fax: (972-2)388704
E-Mail: hydrolog@vms.huji.ac.il

Institute for Petroleum Research and Geophysics (G,R)
1 Hamashbir Street
P.O. Box 2286
Holon 58122
Director General: Y. Rotstein
Phone: (972-3) 5576050
Fax: (972-3) 5502925
E-Mail: yair@iprg.energy.gov.il

Israel Geological Society
Box 1238
Jerusalem
Yaacov Arkin
Phone: (972-2) 314259
Fax: (972-2) 380688
E-Mail: arkin@ilgsi

Oceanographic & Limnological Research Institute (H,E)
Tel Shikmona
P.O. Box 8030
Haifa 31080
Director General: Dr. Yuval Cohen
Phone: (972-4)8515202
Fax: (972-4)8511911
E-Mail: yuval@ocean.org.il

Tahal-Water Planning for Israel and Tahal Consulting Engineers, Ltd. (E,G,H)
54 Ibn Gvirol Street
P.O. Box 11170
Tel Aviv 61111
President: Y. Ben-Gal
Phone: (972-3)6924434
Fax: (972-3)6969969/6968818
Telex: 33654

Tel-Aviv University
Department of Geophysics and Planetary Sciences
Ramat Aviv, University Campus
Ramat Aviv
Phone: 03-64081111

ITALY

Association of Italian Geophysicists
Associazione Geofisica Italiana
CNR-IFA
Piazzale Luigi Sturzo
31 00144 Rome
A. Longhetto
Phone: (39-6) 30-483-950
Fax: (39-6) 30-483-055

Central Hydrographic Service (H)
Ministero dei Lavori Pubblici
Via Nomentana 2
00161 Roma
Director: Ing. Sergio Dall'Oglio
Phone: (39-6)8482
Telex: 622314

Department of Geophysics and Volcanology
Dipartimento di Geofisica e Volcanologia
Largo S. Marcellino 10
I-80138 Naples

European Seismological Commission
International Association of Seismology and Physics of the Earth's Interior
Univerisita degli Studi di Trieste, Istituto di Minere e Geofisica Applicata
I-34126 Trieste
C. Morelli

Geological Survey of Italy (G)
Presidenza del Consiglio dei Ministri
Dipartimento per i Servizi Tecnici Nazionali
Roma 00185
Director: Andrea Todisco
Phone: (39-6)4464929
Fax: (39-6)4465159

International Institute for Geothermal Research (G)
Piazza Solferino 2
56120 Pisa
Director: Dr. Paolo Squarci
Phone: (39-50)41503
Fax: (39-50)47055
Telex: 502020

International Institute of Volcanology (G)
Istituto Internazionale di Vulcanologia
Piazza Roma, 2
95123 - Catania
Director: Litterio Villari
Phone: (39-95)448084
Fax: (39-95)435801
http://www.iiv.ct.cnr.it

Italian Geological Society
Societa Geologica Italiana
Departimento de Scienze della Terra
Citta Universitaria
00185 Rome
Phone: (39-6) 4959390

Italian Geotechnical Society
Associazione Geotechnica Italiana
Via Bormida n. 2
00198 Rome
G. Baldi
Phone: (39-6) 8416120
Fax: (39-6) 8842265

Italian Glaciological Committee
Comitato Glaciologico Italiano
Via Accademia delle Scienze 5
I-10123 Torino
Augusto Biancotti
Phone: (39) 6502663

Italian Group of A.I.P.E.A.
CNR-IRTEC
Via Granarolo, 64
48018 Faenza
Bruno Fabbri
Phone: (39-546) 46147
Fax: (39-546) 46381
E-Mail: fabbri@irtec1.irtec.bo.cnr.it

Italian Military Geographic Institute (C,G)
Via Cesare Battisti 10
50122 Firenze
Director: Giancarlo Fabbri
Phone: (39-6)055 2776222
Fax: (39-6)055 282172
http://www.nettuno.it/fiera/igmi/igmit.htm

Italian Paleontological Society
Societa Paleontologica Italiana
Istituto di Paleontologia, via Universita N. 4
41100 Modena
Davoli Franco
Phone: (39-59) 217084
Fax: (39-59) 218212
E-Mail: serpagli@c220.unimo.it

Navy Hydrographic Institute (H)
Passo Osservatorio 4
161134 Genoa
Director: Capt. Giuseppe Angrisano
Phone: (39-10)265451
Fax: (39-10)265973
Telex: 275521

Seismic Service (G)
Ministero dei Lavori Pubblici
Via Nomentana 2
00161 Roma
Acting Director: Ing. Attilio Cipollini
Phone: (39-6)8482
Telex: 622314

Third World Academy of Sciences
International Centre for Theoretical Physics
Box 586
I-34100 Trieste
Abdus Salam
Phone: 40 2240-1
Fax: 40 224559

Vesuvian Vulcanological Observatory (G)
Via Osservatorio
80056 Ercolano (Napoli)
Director: Prof. Paolo Gasparini
Phone: (39-81)7390644
Fax: (39-81)5523596
Telex: 710306

JAMAICA

Geological Society of Jamaica
Department of Geology
University of the West Indies
Mona, Kingston 7
Michael Blackwood
Phone: (809) 927-2728
Fax: (809) 927-1640

Institute of the Expanding Earth (G)
Box 144
Kingston 5
Richard W. Guy
Phone: (809) 977-5999
Fax: (809) 929-1096

Ministry of Agriculture and Mining (G)
Mines and Geology Division
P.O. Box 141
Hope Gardens
Kingston 6
Commissioner: Coy Roache

Phone: (809)927-1936/40
Fax: (809)977-1204

National Water Commission (H)
P.O. Box 65
14-19 Trinidad Terrace
Kingston 5
Phone: (809)9295430/5

Scientific Research Council
Box 350
Kingston 6
Merlene E. Bardowell
Phone: (809) 92-71771-4
Fax: (809) 92-75347

Survey Department (C)
P.O. Box 493
231/2 Charles Street
Kingston
Director: L.R. Bulgin
Phone: (809)9226630/5
Fax: (809)9671010

Water Resources Authority (H)
Ministry of Public Utilities and Transport
P.O. Box 91, Hope Gardens
Kingston 7
Managing Director: Basil P. Fernandez
Phone: (809)927-0077
Fax: (809)977-0179

University of the West Indies
Geology Department
Mona
Kingston 6
Phone: (809)927-2728
Fax: (809)927-1640

JAPAN

Agency of Natural Resources and Energy (G)
Ministry of International Trade and Industry (MITI)
1-3 Kasumigaseki
Chiyoda-ku, Tokyo 100
Director General: Teiichi Yamamoto
Phone: (81-3)5011511

Aichi University of Education
Department of Earth Sciences
Igaya
Kariya
Shizuoka 448

Akita University
Department of Mining Geology
Tegata
Akita 010

Chiba University
Department of Earth Sciences
Yayoi
Chiba 260

Doshisha University
Laboratory of Earth Sciences
Kamikyo
Kyoto 606

Earthquake Research Institute (G)
University of Tokyo
1-1 Yayoi-cho 1-chome
Bunkyo-ku, Tokyo 113
Director: Yoshio Mogi
Phone: (81-3)8122111
Telex: 2722148 ERI J

Ehime University
Department of Earth Sciences
Bunkyo
Matsuyama 790

Fukui University
Department of Earth Sciences
Bunkyo
Fukui 910

Fukuoka University of Education
Department of Earth Sciences
Akama
Munakata
Fukuoka Prefecture 811-41

Geochemical Society of Japan
Nihon Chikyo Kagakkai
Business Center for Academic Societies
16-9 Honkomagome 5-chome
Bunkyo-ku
Tokyo 113
Nobuo Takaoka
Phone: (81-92) 642-2668
Fax: (81-92) 642-2684
E-Mail: takaoka@geo.kyushu-u.ac.jp

Geographical Survey Institute (C)
Ministry of Construction
1, Kitasato, Tsukuba
Ibaraki 305
Director General: Kunio Nonomura
Phone: (81-298)641111
http://www.gsi-mc.go.jp/

Geological Society of Japan
Maruishi Building/10-4
Kajicho 1 chome
Chiyoda-ku
Tokyo 101

President: Masahiko Akiyama
Phone: (03) 3252-7242
Fax: (03) 5256-5676

Geological Survey of Japan (GSJ) (G,H)
1-1-3 Higashi
Tsukuba, Ibaraki 305
Director General: Hirokazu Hase
Phone: (81-298) 54 3576
Fax: (81-298) 56 4989
E-Mail: hase@gsj.go.jp
http://www.gsj.go.jp

Gifu University
Laboratory of Earth Sciences
Yanagido
Gifu 501-11

Gunma University
Department of Earth Sciences
Aramaki
Maebashi
Gunma
Gunma 371

Himeji Institute of Technology
Laboratory of Earth Sciences
Shosha
Himeji 671-22

Hirosaki University
Department of Earth Sciences
Bunkyo
Hirosaki
Aomori 036

Hiroshima University

Department of Geology and Mineralogy
Higashisenda
Naka-ku
Hiroshima 730

Laboratory of Natural Environment
Higashisenda
Naka-ku
Hiroshima 730

Hokkaido University of Education
Department of Earth Sciences
Ainosto
Kita-Ku
Sapporo 002

Hokkaido University
Department of Geology and Minerology
Kita 8-Chome
Kita-Ku
Sapporo 060

Hyogo Kyoiku University
Institute of Natural Sciences
Yashiro-cho
Kato-gun
Hyogo 673-14

Ibaraki University
Department of Earth Sciences
Bunkyo
Mito 310

College of Integrated Arts and Sciences
Laboratory of Earth Sciences
Mozume
Sakai
Osaka 591

International Paleontological Association
Department of Geology and Mineralogy
Faculty of Science
Hokkaido University
Sapporo 060
Mokoto Kato
Phone: (81-11) 716 2111

Iwate University
Department of Earth Sciences
Uyeda
Morioka
Iwate 020

Japanese Association of Mineralogists, Petrologists and Economic Geologists
c/o Faculty of Science
Tohoku University, Aoba
Sendai 980
President: Hitoshi Onuki
Phone: (81-22) 224-3852
Fax: (81-22) 224-3852
E-Mail: kyl04223@niftyserve.or.jp

Japanese National Committee of Geology
Science Council of Japan
22-34 Roppongi 7-chome
Minato-ku
Tokyo 106
Tadashi Sato
Phone: (81-3)3403-6291
Fax: (81-3)3403-6224

Kagawa University
Department of Earth Sciences
Saiwai-cho
Takamatsu 760

Kagoshima University
Department of Earth Sciences

Gungen
Kagoshima 890

Kanazawa University
Department of Earth Sciences
Marunouchi
Kanazawa 920

Kobe University
Department of Earth Sciences
Rokkodai
Nada
Kobe 657

Kochi University
Department of Geology
Akebono-cho
Kochi 780

Kumamoto University
Department of Geology
Kurokami
Kumamoto 860

Kyoto University
Department of Geology and Mineralogy
Kitashirakawa
Sakyo
Kyoto 606

Kyoto University
Disaster Prevention Research
Gokanosho
Uji
Kyoto 611

Kyoto University
  Department of Mineral Science and
    Technology
  Yoshida
  Sakyo
  Kyoto 606

  Institute of Earth Science
  Yoshida
  Sakyo
  Kyoto 606

Kyushu University
  Department of Earth and Planetary
    Sciences
  Hakozaki
  Higshi-ku
  Fukuoka 812

  Department of Mining
  Hakozaki
  Higshi-ku
  Fukuoka 812

Meijo University
Laboratory of Geoscience
Shiogamaguchi
Tempaku
Nagoya 468

Mie University
Department of Earth Sciences
Kamihama Tsu
Mie 514

Mineralogical Society of Japan
Nogizaka Building
Akasaka 9-6-41
Minato-ku
Tokyo 107
Hiroyuki Horiuchi
Phone: (81-3)3812-2111 EXT. 4542
Fax: (81-3)3812-5714
E-Mail: horiuchi@s.u.-tokyo.ac.jp

Miyagi University of Education
Department of Earth Sciences
Aramaki
Aoba-Ku
Sendai
Sendai 980

Miyazaki University
Laboratory of Earth Sciences
Konohanadai
Miyazaki 880

Muroran Institute of Technology
Department of Geotechnology
Mizumoto
Muroran 050

Nagasaki University
Department of Earth Sciences
Bunkyo
Nagasaki 852

Nagoya University
  Department of Earth Sciences
  Furo-Cho
  Chigusa
  Nagoya 466

  Radioisotope Center
  Furo-Cho
  Chigusa
  Nagoya 466

  Water Research Institute
  Furo-cho
  Chigusa
  Nagoya 466

Nara University of Education
Department of Earth Sciences
Takabatake
Nara 630

National Institute for Environmental Studies (E,H)
Onogawa 16-2, Tsukuba
Ibaraki 305
Director General: Tsuguyoshi Suzuki
Phone: (81-298)50-2308
Fax: (81-298)51-2854
http://www.nies.go.jp

National Research Institute for Resources and Environment (E,G)
16-3 Onogawa, Tsukuba
Ibaraki 305
Director: Nagayuki Yokoyama
Phone: (81-298)543000
Telex: 365-2570 AIST J

Nihon University
Department of Applied Earth Sciences
Sakuragami
Setagaya
Tokyo 156

Niigata University
Department of Earth Sciences
Nino-machi
Igarshi
Niigata 950-21

Department of Geology and Mineralogy
Nino-machi
Igarshi
Niigata 950-21

Oita Univesity
Department of Earth Sciences
Tannobaru
Oita 870-11

Okayama University of Science
Nirusen Research Institute
Kawakami
Maniwa
Okayama Prefecture 717-06

Okayama University of Science
Department of Fundamental Natural Science
Ridai-Cho
Okayama 700

Okayama University
Department of Earth Sciences
Naka
Tsushima
Okayama 700

Institute of Study of the Earth's Interior
Misasa
Tottori Prefecture 682-02

Osaka City University
Department of Geosciences
Sugimoto
Sumiyoshi
Osaka 558

Osaka Institute of Technology
Laboratory of Applied Geology
Osaka

Osaka Kyoiku University (Ikeda Campus)
Department of Earth Sciences
Jonan
Ikeda
Osaka 558

Osaka Kyoiku University (Tennoji Campus)
Department of Earth Sciences
Minami-Kawabori
Tennoji
Osaka 543

Osaka University
Department of Earth Sciences
Machikaneyama
Toyonaka
Osaka 560

Palaeontological Society of Japan
c/o Business Center for Academic Societies
Yayoi 204-6
Bunkyo-ku
Tokyo 113
Kiyotaka Chinzei

Public Works Research Institute (G)
Ministry of Construction (includes engineering geology)
1, Asahi, Oaza, Tsukuba-shi
Ibaraki 305
Director: Toshio Iwasaki
Phone: (81-298)642211

Saga University
Laboratory of Earth Sciences
Honjo
Saga 840

Saitama University
Department of Natural Sciences
Shimo-Okubo
Urawa
Saitama 338

Seismological Society of Japan
Nippon Zisin Gakkai
Earthquake Research Institute
University of Tokyo

1-1, Yayoi 1-chome, Bunkyo-ku
Tokyo 113
Yoshio Fukao
Fax: (81-3) 5684-2549
E-Mail: zisin@eri.u-tokyo.ac.jp

Shiga University

Department of Earth Sciences
Hiratsu
Otsu 520

Shimane University

Department of Gelogy
Nishikawazu
Shimane 690

Shinshu University

Department of Geology
Asahi
Matsumoto 390

Shizuoka University

Department of Geosciences
Otani
Shizuoka 422

Society of Resource Geology

Shigen-Chishitsu-Gakkai
Nogizaka Building 3F
Akasaka 9-6-41
Minato-ku
Tokyo 107
Hidehiko Shimazaki
Phone: (81-3) 3475-5287
Fax: (81-3) 3475-0824

Tohoku University

Department of Geology and Paleontology
Aramaki
Aoba-Ku
Sendai 980

Department of Mineralogy, Petrology and Economic Geology
Aramaki
Aoba-Ku
Sendai
Sendai 980

Department of Mining and Mineral Geology
Aramaki
Aoba-Ku
Sendai 980

Tokai University

Institute of Marine Resources
Orido
Shimizu
Shizuoka 424

Tokushima University

School of Synthetic Science
Tsunemishima
Tokushima 770

University of Tokyo

Earthquake Research Institute
Yayoi
Bunkyo
Tokyo 113

University of Tokyo

Department of Earth and Astronomy
Komaba
Meguro
Tokyo 153

University of Tokyo

Department of Geology
Hongo
Tokyo 113

University Museum
Hongo
Tokyo 113

University of Tokyo

Ocean Research Institute
Mimamidai
Nakano
Tokyo 164

Tokyo Geographical Society

Tokyo Chigaku Kyokai
12-2 Nibancho
Chiyoda-ku
Tokyo 102
Yoshiro Ueda
Phone: (81-3) 3261-0809
Fax: (81-3) 3263-0257

Tokyo Institute of Technology

Department of Earth Sciences
Ookayama
Meguro
Tokyo 152

Tottori University

Department of Earth Sciences
Kozan
Tottori 680

Toyama University

Department of Earth Sciences
Gofuku
Toyama 930

Toyo University

Natural Science Laboratory
Oka
Asaka 351

University of Tsukuba

Environmental Research Center
Tennodai
Tsukuba 305

Institute of Geoscience
Tennodai
Tsukuba 305

Volcanological Society of Japan

Nippon Kazan Gakkai
Earthquake Research Institute
University of Tokyo
1 Yayoi-cho, Bunkyo-ku
Tokyo
Phone: (81-3) 813-7421
Fax: (81-3) 5684-2549

Waseda University

Department of Resources Technology
Okubo
Shinjuku
Tokyo 169

Water Resources Bureau (H)

National Land Agency
Prime Minister's Office
1-2-2 Kasumigaseki, Chiyoda-ku
Tokyo 100
Director General: Mitsuro Ohkawara
Phone: (81-3)5933311

Yamagata University

Departement of Earth Sciences
Koshirakawa
Yamagata 990

Yamaguchi University

Department of Earth Sciences
Yoshida
Yamaguchi 735

Department of Mineralogical Sciences and Technology
Yoshida
Yamaguchi 735

Yamanashi University

Department of Earth Sciences
Takeda
Kofu 400

Yokohama National University

Department of Earth Sciences
Tokiwadai
Hodogaya
Yokohama 240

JORDAN

Department of Lands and Surveys (C)

Ministry of Finance
P.O. Box 70
Amman
Director: Badri Al-Mulqi

Natural Resources Authority (NRA) (C,G,H)

P.O. Box 7
Amman
Director General: Khalid Sheyyab
Phone: (962-6)811808
Fax: (962-6)811866

Royal Jordanian Geographic Centre (C)

P.O. Box 20214
Amman - 11118
Director: Saliem Khalifa
Phone: (962-6)845188
Fax: (962-6)847694

Royal Scientific Society (G)

P.O. Box 6945
Amman
President: Albert Butros

KAZAKHSTAN

Research Institute of Mineral Resources (G)

K Marx St., 105
Almaty 480091
Director: Ginajat Bekzhanov
Phone: (7-3272)612076

KENYA

Department of Geology (G)

University of Nairobi
P.O. Box 30197
Nairobi
Chairman: Dr. I.O. Nyambok

Geological Society of Africa

Box 60199
Nairobi
Frederick Githui Theuri
Phone: (254-2) 541040

Geological Survey of Kenya (G)

P.O. Box 30750

Kenya

Nairobi
Chief Geologist: A.E. Johnson

Mines and Geological Department
Ministry of Environment and Natural
 Resources
Headquarters, Kencom House
City Hall Way/Moi Ave.
Box 30009
Nairobi
C.Y.O. Owayo
Phone: (254-2) 229261

Regional Center for Services in Surveying, Mapping and Remote Sensing (C,E)
P.O. Box 18118
Nairobi
Director: Simon L.P. Ndyetabula
Phone: (254-2)803320/9
Fax: (254-2)802767
E-Mail: ndyetab@ressmrs.rio.org

Survey of Kenya (C)
P.O. Box 30046
Nairobi
Director of Surveys: D. Kamau

United Nations Environment Programme
Box 30552
Nairobi
Elizabeth Dowdeswell
Phone: (254-2)621234
Fax: (254-2)226890

Water Department (H)
Ministry of Water Development
P.O. Box 30521
Nairobi
Water Eng. Dept.: C.N. Mutitu
Water Res. Dept.: D.N. Kirori

KIRIBATI

Ministry of Local Government and Rural Development (G)
Lands and Survey Division
P.O. Box 405
Betio, Tarawa

Ministry of Natural Resources Development (G)
P. O. Box 64
Bairiki, Tarawa
Secretary for Natural Resources Development: Tinian Reiher
Phone: (686) 21099
Fax: (686) 21120
E-Mail: tinianr@mnrd.ki.gov

KOREA, DEMOCRATIC PEOPLE'S REPUBLIC OF

Hydro-Meteorological Bureau (H)
P.O. Box 100
P'yongyang
Director: Li Gon Il

Institute of Geology (C)
State Academy of Sciences
Kwahak-1Dong, Unjong District
P'yongyang
Director: Kang Hyong Gap
Fax: 850-2-3814580

KOREA, REPUBLIC OF

Bureau of Water Resources (H)
Ministry of Construction and Transportation
The 2nd Unified Government Building
1, Choongang-dong, Kwacheon City
Kyounggi-do
Director General: Kim, Chang-se
Phone: (82)5039040
Fax: (82)5037409
Telex: K24755

Geology and Mineral Institute of Korea
219-5 Karibong-dong
Yongdungpo-ku
Seoul

Korea Institute of Geology, Mining and Materials (KIGAM) (C,G)
30 Kajungdong, Yusongku
Taejon 305-350
President: Dr. Dong-Hak Kim
Phone: (82-421)8683270
Fax: (82-421)8619720
Telex: KIERSK K45509

Korea Ocean Development Research Institute (KORDI) (H)
385 Bloc 4-dong
Ansan City, Kyonggido
President: Won-Oh Song
Phone: (82-345)8634770
Fax: (82-345)826698
Telex: K27675

LITHUANIA

Geological Survey of Lithuania (E,G,R)
Konarskio 35
Vilnius 2600
Director: Gediminas Motuza
Phone: (370-2) 632889
Fax: (370-67)06376
E-Mail: gediminas.motuza@lgt.lt

JSC Vilnius Hydrogeology Ltd. (G)
Basanaviciaus 37/1
Vilnius 2009
Chief Hydrogeologist: A. Algirdas Klimas
Phone: (012-2)650156
Fax: (012-2)235058

LUXEMBOURG

Administration du Cadastre et de la Topographie (C)
54, Avenue Gaston Diderich
B.P. 1761
L-1017 Luxembourg
Director: AndreMajerus
Phone: (352)44901-1
Fax: (352)44901-288

Centre Européen de Géodynamique et de Seismologie (C,G)
Rue Josy Welter, 19
L-7256 Walferdange
Nicholas d'Oreye
Phone: (352)336130
Fax: (352)336129

Service de la Meteorologie et de l'Hydrologie (H)
16, route d'Esch
B.P. 1904
L-1019 Luxembourg
Director: R. Kipgen
Phone: (352)45 71 72 - 326
Fax: (352)45 71 72 - 341

Service Geologique (C,G)
Ponts et Chaussees
43. Bld. G.D.-Charlotte
L-1331 Luxembourg
Director: Robert Maquil
Phone: (352)444126
Fax: (352)458760
E-Mail: robert.maquil@life.lu

MACEDONIA, REPUBLIC OF

University "St. Cyril and Methodius" (G)
Institute of Earthquake Engineering and Engineering Seismology
Salvador Aljende Str. 73
P.O. Box 101
91000 Skopje
Director General: Dimitar Jurukovski
Phone: (389-91)112154
Fax: (389-91)112163
E-Mail: institut@plato.iziis.ukim.edu.mk

MADAGASCAR

Centre National de Recherches Oceanographiques (G)
Ministere de la Recherche Scientifique et Technologique pour le Developement
B.P. 68
207 Nosy Be
Director: Guy Arthur Andriamirado Rabarison
Phone: (261-2)613373 Nosy Be
Telex: 22539 MRSTD MG ANTANANARIVO

Institut Géographique et Hydrographique National (C,H)
Rue Dama-Ntsoha Razafintsalama Jean Baptiste
P.O. Box 323 - Ambanidia
101 - Antananarivo
General Manager: Naina Andriamparany Aimé
Phone: (261-2)22935
Fax: (261-2)25264
E-Mail: ftm@bow.dts.mg

Ministere de l'Industrie, de l'Energie et de Mines (MIEM) (C,G,H)
B.P. 257
101 Antananarivo
Director: M. Jaona Andriansolo
Phone: (261-2)25515
Telex: 098622540 MG

Service des Mines et de la Geologie (G)
B.P. 280
101 Antananarivo
Director: Jaona Andrianasolo
Phone: (261-2)40191
Telex: 22540 MIEM MG

MALAWI

Department of Mines (R)
P.O. Box 251
Lilongwe
Acting Chief Mining Engineer: Y.R. Phiri
Phone: (265)722933
Fax: (265)722772
Telex: 43159 MINS MI

Department of Surveys (C,R)
Ministry of Physical Planning and Surveys
P.O. Box 349
Blantyre
Surveyor General: A.F. Tambala
Phone: (265)623722
Fax: (265)634034

Geological Survey Department (G,H)
Ministry of Energy and Mining
P.O. Box 27, Liwonde Road
Zomba
Chief Geologist: P.R. Phiri
Phone: (265-50)522166/522670
Fax: (265-50)522716
Telex: 44382 GEOLOGY MI

Ministry of Research and Environmental Affairs (E)
Office of the President and Cabinet
P.O. Box 30745
Lilongwe
Environmental Co-ordinator: R. P. Kabwadza
Phone: (265)781111
Fax: (265)781487
Telex: 45311 DREA MI
E-Mail: rkabwadza@unima.wn.apc.org

Water Resources Division (H)
Ministry of Works
P.O. Box 30026
Lilongwe 3
Chief Water Resources Officer: R.D. Kafundu
Phone: (265)732155
Telex: 44285 Work MI

MALAYSIA

Department of Irrigation and Drainage, Malaysia (H)
Ministry of Agriculture
Jalan Sultan Salahuddin
50626 Kuala Lumpur
Director General: Neo Tong Lee
Phone: (60-3)2982618
Fax: (60-3)2914282
Telex: MA 33045

Directorate of National Mapping (C)
Bangunan Ukor, 1st Floor
Jalan Gurney
50578 Kuala Lumpur
Director: Encik Abdul Majid bin Mohamed
Phone: (60-3)2925311, Ext. 291
Fax: (60-3)2917457
Telex: MA 28148

Geological Society of Malaysia
Persatuan Geologi Malaysia
d/a Jabatan Geologi
Universiti Malaya
50603 Kuala Lumpur
Phone: (60-3) 7577036
Fax: (60-3) 7563900

Geological Survey of Malaysia (G)
Tabung Haji Building
19-21 Floor
Jalan Tun Razak
P.O. Box 11110
50736 Kuala Lumpur
Director General: Fateh Chand
Phone: (60-3)2611033/1034/1035/1325
Fax: (60-03)2611036
Telex: MA 30808
E-Mail: kbip@gsmkl1.gsm.gov.my

University of Malaya
Department of Geology
Kuala Lumpur 50603
Phone: (6) 03 8250001/10
Fax: (6) 03 8256086

National University of Malaysia
Department of Geology
Faculty of Physical and Applied Sciences
Bangi, Selangor D.E.
43600

Petroliam Nasional Berhad (PETRONAS) (R)
P.O. Box 12444
50778 Kuala Lumpur
Phone: (60-3)2743833/8011/8022
Fax: (60-3)2740217
Telex: PETRON MA 31123

Universiti Sains Malaysia
Physical Science Study Centre
Pulau Pinang 11800

MALI

Direction Nationale des Mines et de la Geologie (G)
B.P. 223
Koulouba, Bamako
Director General: M. Sekou Diallo

Institut National de Topographie (C)
B.P. 240
Bamako
Director: Diadie Traore

Service Hydrologique (H)
Direction Generale de l'Hydraulique et de l'Energie
B.P. 66
Bamako
Chief: Cisse Navon
Phone: (223)22 4877
Fax: (223)22 8635

Societe Nationale de Recherches et d'Exploitation Miniere (SONAREM)
B.P. 2
Kati
Phone: (223) 27-20-49
Fax: (223) 27-20-42

MALTA

Hydrology Section (H)
Water Works Department
Beltissebh
Floriana
Chief: Dr. Godwin Debono
Phone: (356)225586
Telex: 110 MOD MLT MT

Works Department (G)
Project House
Floriana
CMR 02
Director General: Vincent Cassar B. Arch
Phone: (356)221410/221411
Fax: (356)234145
E-Mail: worksdiv@maltanet.omnes.net

MARSHALL ISLANDS

Marshall Islands Marine Resources Authority (MIMRA) (G,R)
P.O. Box 860
Director: Danny S. Wase
Phone: (60)6928262
Fax: (60)692625/5447

RMI Environmental Protection Authority (E)
P.O. Box 1322
Majuro MH 96960
Director: Jorelik Tibon
Phone: (60)6926255203
Fax: (60)6926255202

MARTINIQUE

Observatoire Volcanologique de la Montagne Pelee
Fonds Saint Denis
97250 Saint Pierre

Office of Overseas Scientific and Technical Research (G,H)
B.P. 81
97000 Fort-de-France
Director: Jean-Alfred Gueredrat
Phone: (596)702872

MAURITANIA

Direction de la Cartographie et de la Topographie (DCT) (C)
B.P. 237
Nouakchott
Director: Mohamed Ould Brahim
Phone: (222-2)55226

Direction de la Protection de la Nature (DPN) (E)
B.P. 170
Nouakchott
Director: Dahmoud Ould Merzoug
Phone: (222-2)51834
Director: Ely Ould El Haj
Phone: (222-2)57140
Fax: (222-2)51402

Direction de l'Hydraulique (DH) (H)
Director: Ely Ould El Haj
Phone: (222-2) 57140
Fax: (222-2) 51402

Ministere des Mines et de la Geologie (G,R)
B.P. 199
Nouakchott
Minister: M'Boye Ould Arafa
Phone: (222-2)53225

Office Mauritanien de Recherches Geologiques (OMRG) (G)
B.P. 654
Nouakchott
Director: Ishaq Ould Ragel
Phone: (222-2)51410

MAURITIUS

Central Water Authority (H)
Saint Paul
Phoenix
Phone: (09230)686-5071
Fax: (09230)686-6264

University of Mauritius
Faculty of Engineering
Reduit
Phone: (09230)454-1041
Fax: (09230)454-9642

Mauritius Sugar Industry Research Institute (G,R)
Reduit
Director: R. Julien
Phone: (09230)454-1061
Fax: (09230)454-1971

Ministry of Housing, Lands, and Town and Country Planning (C)
Survey Department
Edith Cavell Street
Port Louis
Chief Surveyor: D. Ramasawmy
Phone: (09230)208-2831
Fax: (09230)212-9369

Ministry of the Environment and Quality of Life (E)
Corner of St. George and Barracks Streets
Port Louis
Director of Environment: R. Prayag
Phone: (09230)212-8332
Fax: (09230)212-9407

MEXICO

Universidad Autonoma de Guerrero
Facultad de Ciencias de la Tierra
Texco, El Viejo
Av. Aleman , Col. Cuauhtemoc Sur
Chilpancingo, Gro. 39060

Universidad Autonoma de Nuevo Léon
Departamento de Geohidrologia
Ciudad Universitaria
Espinoza y Ruperto Martinez
San Nicolas de los Garza
N.L. 64000

Facultad de Ciencias de la Tierra
Hda de Guadalupe, Camino Cerro Prieto Km8
A.P. 104 - Linares, N.L. 67700

Universidad Autonoma de Yucatan
Ingenieria Ambiental
Calles 114 y 41
Merida, Yucatan

Universidad Autonoma de Zacatecas
Escuela de Ingenieria de Minas, Metalurgia y Geologia
Calzada de la Universidad s/n 98000

Centro de Investigacion Cientifica
Educacíon Superior de Ensenada
División de Ciencias de la Tierra
Carr. Tijuana-Ensenada km 197
Ensenada, Baja California Norte

Colegio de Ingenieros Geólogos de México, A.C.
Av. Insurgentes Sur 949, 4 piso
Col. Nápoles
Ciudad de México
C.P. 03810 México, D.F.
Phone: (52-5) 543-9782
Fax: (52-5) 543-8254

Comisíon Federal de Electricidad
Gerencia de Ingenieria Civil
Oklahoma No. 85 Col. Napoles
Mexico, D.F. 03810

Comisíon Nacional del Agua
Av. Insurgentes Sur 2140, 2 Piso Col. Ermita
Mexico, D.F. 01070

Consejo de Recursos Minerales
Boulevard Felipe Angeles s/n
Carretera Mexico-Pachuca Km 93 5 Col. Venta Prieta
Pachuca, Hgo. 42080

Consejo Nacional de Ciencia y Tecnologia
Circuito Cultural Centro Cultural Universitario
Ciudad Universitaria
Mexico, D.F. 04515

Universidad de Guadalajara
Centro de Ciencias de la Tierra

Juan N Cumplido
Sector Hidalgo
Guadalajara, Jal. 44100

Departmento Geografico Militar (C,G)
Servicio Cartografico
Secretaria de la Defensa Nacional
Lomas de Sotelo
Mexico 10, D.F.
Director General: General de Brigada Agustin Zarate Guerrero

I.P.N. - E.S.I.A.
Ciencias de la Tierra
Av. Ticoman 0
Col. San Jose Ticoman
Mexico, D.F. 07330

Instituto de Geologia
Estacion Regional del Noroeste
Universidad Nacional Autonoma de Mexico
Apartado Postal 1039
83000 Hermosillo, Sonora
Phone: (52-62) 175019
Fax: (52-62) 175340
E-Mail: cmgleon@servidor.unam.mx

Instituto de Investigaciones Electicas
Dep. de Geotermia
A.P. 475 Centro
Cuernnavaca, Morelos 62000

Instituto Mexicano de Tecnologia del Agua
Paseo Cuahunahuac 8532
Jiutepec
A.P. 235, Civac. Morelos

Instituto Mexicano del Petroleo
Eje Central Lazaro Cardenas
152 Col. San Bartolo
Atepehuacan, Mexico D.F. 07730

Instituto Mexicano del Petroleo (G)
Av. Cien Metros No. 152
Mexico 14, D.F.
Director General: Francisco Barnés de Castro
Phone: (52-5)567-2962
Fax: (52-5)368-9399
http://www.imp.mx

Instituto Nacional de Investigacíon Nuclear
Centro Nuclear Salazar
Sierra Mojada 447-201
Lomas de Barirlaco
Mexico, D.F. 10010

Instituto Tecnologico de Cd. Madero
Departamento de Geologia

1 de Mayo y Sor Juana Ines de la Cruz s/n
Cd. Madero, Tamaulipas C.P. 89400

Mexican Association of Exploration Geophysicists
Asociacion Mexicana de Geofisicos de Exploracion
Apartado Postal 57-275
C.P. 06500
Raymundo Aguilera Ibarra
Phone: (52) 531-63-08
Fax: (52) 250-26-11, EXT 26812

Mexican Association of Petroleum Geologists
Asociacion Mexicana de Geologos Petroleros
Torres Bodet 176
06400 Mexico, D.F.

Mexican Geological Society
Sociedad Geologica Mexicana A.C.
Torres Bodet No. 176
Col. Santa Maria La Ribera
C.P. 06400
Bernardo Martell Andrade
Phone: (52) 547-26-66

Mexican Geophysical Union
Union Geofisica Mexicana
Apartado Postal 2732
Ensenada 22800 B.C.
M.C. Francisco Suavez V.

Universidad Nacional Autonoma de Mexico (UNAM) (G)
Instituto Geofísica
Ciudad Universitaria
A.P. 21-499
Mexico 20, D.F.
Gerardo Suárez
Phone: (52-5)550-6662
E-Mail: gerardo@ollin.igeofcu.unam.mx

Secretaria de Agricultura, Ganadería y Desarrollo Rural (H)
Avenida Insurgentes Sur No. 476, piso 13
Col. Roma Sur
06760 Mexico, D.F.
Secretary General: Francisco Labastida Ochoa
Phone: 584-00-96, 584-02-71
Fax: 584-02-68

Secretaria de Medio Ambiente (E)
Recursos Naturales y Pesca
Lateral de Periférico Sur, 4209
Fraccionamiento Jardines de la Montaña
CP 14210, México, D.F.
Secretary General: Julia Carabias
Phone: (525-6) 28 06 02
Fax: (525-6) 28 06 03

MICRONESIA, FEDERATED STATES OF

Department of Resources and Development (C,G,H)
P.O. Box PS-12
Palikir, Pohnpei 96941
Secretary: Sebastian L. Anefal
Phone: (691) 320-2646
Fax: (691) 320-5854
Telex: 7296807 FSMGOV FM

MONGOLIA

Ministry of Geology and Mineral Resources (E,G,H)
P.O.B. 37/110
Ulaanbaatar - 37
Director: Namjiliin Jadambaa
Phone: (976-1)33-28-95
Fax: (976-1)33-10-84

Geological Survey of Mongolia
P.O. Box 91, 37 Tolgoit
Ulaanbaatar
General Director: Dr. Jambiin Byamba
Phone: (976-1)332895
Telex: 79308 MGGE MH

Mongolian Geological and Mining Society
Mongolian Geological and Geophysical Exploration Co.
Peace Av 47
Ulaanbaatar 51
210351 Mongolia
Ulaanbaatar - 32
c/o D. Bat-Erdene

MOROCCO

Association of African Geological Surveys
Institut de Geolgie
Nouveau Quarier Administratif
Agdal-Instituts
B.P. 6208
Rabat
M. Beisaid

Centre Royal de Teledetection Spatiale (G)
16 bis, Ave. de France
Agdal-Rabat, Maroc
Phone: (212-7)776305
Fax: (212-7)776300
Telex: 31 761 M

Geological Survey of Morocco (G)
Ministere de l'Energie et des Mines
Quartier Administratif
Rabat
Director: Dr. Mohamed BenSaid
Phone: (212-7)72824
Telex: 32761M

Ministère de L'Agriculture et de la Mise en Valeur Agricole (C)
Administration de la Conservation Foncière du Cadastre et de la Cartographie
Avenue Moulay Youssef
Rabat
Director: Abdelatif Belbachir
Phone: (212-7)70-90-01
Fax: (212-7)70-58-85
Telex: 31038M

Ministere de l'Agriculture et de la Reforme Agraire
Conservation Fonciere et des Travaux Topographiques
Avenue Moulay Youssef
Rabat
Director: Abdellatif Bel Bachir
Phone: (212-7)65717
Telex: 31038M

Ministère des Travaux Publics (H)
Direction Générale de l'Hydraulique
rue Hassan Benchekroun, Agdal
Casier Rabat-Chellah
Director: Mohammed Jellali
Phone: (212-7)77 90 08
Fax: (212-7)77 86 96

MOZAMBIQUE

National Directorate for Geography and Mapping (C)
Avenida Josina Machel, 589/Maputo
P.O. Box 288
Maputo
Director: Jose Manuel Almeitim Carvalho
Phone: (258-1)22786
Telex: 6500 DNHMCMO

National Directorate of Geology (C,G)
Praca 25 de Junho, no. 380
P.O. Box 217
Maputo

National Director: Joa Manuel Perdiz Reynolds Marques
Phone: (258-1)420797
Fax: (258-1)429216
Telex: 6584 GEOMI MO
Geologist: Joao M.D. Reynolds Marques

National Directorate of Hydrology (H)
Avenida 25 de Setembro, 942-7th Floor
P.O. Box 1611
Maputo
Director: Policarpo do Rosario Mabica
Phone: (258-1)22191/27011/22834

MYANMAR

Applied Geology Department (G)
6-1/2 Mile Prome Road
Rangoon
Director: Dr. Myint Thein Lwin

Central Research Organization (C)
Kanbe Road, Yankin
Rangoon
Director General: Dr. Maung Maung Gale
Phone: (951-2)50917/50544
Telex: 21503 KASALA BM

Department of Geological Survey and Exploration (G)
Ministry of Mines
Kanbe Road, Yankin P.O.
Rangoon
Managing Director: U Thaung
Phone: (951-2)61902
Telex: 21307 MYCORP BM

Irrigation Department (H)
Ministry of Agriculture
Kanbe Road, Yankin
Yangon
Director General: U Ohn Myint
Phone: (951)578104
Fax: (951)55917
Telex: 21518 IRRIBM

Meteorology and Hydrology Department (H)
Kaba Aye Pagoda Road
Rangoon
Director General: U Ohn Maung
Phone: (951-2)60824
Telex: 21512 BURMET BM

Myanmar Oil Corporation (G)
604 Merchant Street
Rangoon
Managing Director: U Aung Min

Phone: (951-2)81483
Telex: 21307 MYCORP BM

No. (1) Mining Corporation (G)
104 Strand Road
Rangoon
Managing Director: U Yoe Sein
Phone: (951-2)57457
Telex: 21307 MYCORP BM

No. (2) Mining Corporation (G)
Kanbe Road, Yankin P.O.
Rangoon
Managing Director: U Nyan Lin
Phone: (951-2)51421
Telex: 21205 MCTWO BM

No. (3) Mining Corporation (G)
Rangoon-Insein Road, Thamaing
Rangoon
Managing Director: U Myint Thein
Phone: (951-2)57444
Telex: 21205 MCTWO BM

Survey Department (C)
Ministry of Agriculture and Forestry
Thirimingala Lana
Kaba Aye Pagoda Road
Rangoon
Director General: U Maung Maung
Phone: (951-2)62923/62933

NAMIBIA

Geological Society of Namibia
Box 699
Windhoek
GIC Schneider
Phone: (264-61) 249150
Fax: (264-61) 249146
E-Mail: gabi@gsn822.gsn.mme.gov.na

Geological Survey (G)
P.O. Box 2168
Windhoek
Director: G.I. Schneider
Phone: (264-61)249150
Fax: (264-61)249146
Telex: (50)908487 WK

Ministry of Agriculture, Water and Rural Development (H,R)
Department of Water Affairs
Private Bag 13193
Windhoek
Deputy Permanent Secretary: R.G. Fry
Phone: (264-61)3963074
Fax: (264-61)224512

Namibia

Ministry of Mines and Energy (R)
Directorate: Mines
Private Bag 13297
Windhoek
Director: G. Schneider
Phone: (264-61)226571
Fax: (264-61)238643
Telex: 50908487

The Surveyor General (C)
P.O. Box 13182
Windhoek
Phone: (264-61)2859111
Fax: (264-61)228240

NEPAL

Department of Irrigation and Hydrology (H)
Ministry of Water Resources
Panipokhari
Kathmandu
Director General: Mohan Dhoj Karki
Phone: (977)413733

Department of Mines and Geology (G)
Ministry of Industry
Lainchaur
Kathmandu
Director General: Gopal Singh Thapa
Phone: (977-1)414740
Fax: (977-1)414806

Nepal Geological Society
Box 231
Katmandu

Survey Department (C)
Ministry of Land Reform
Dilli Bazar
Kathmandu
Director General: Buddhi Narayan Shrestha
Phone: (977)411897

NETHERLANDS

Cartography Service (C)
Ministry of Defense
Bendienplein 5
7815 SM Emmen
Director: P.W. Geudeke
Phone: (31-70)591 696911
Fax: (31-70)591 696296

European Association of Geoscientists and Engineers
Box 298
3700 AG Zeist
E.H. Bornkamp
Phone: (31) 30 6962655
Fax: (31) 30 6962640
http://www.ruu.nl/eage

Geo Centrum Brabant (G)
St. Lambertusweg 4
5283 VL Boxtel
R.H.B. Fraaye
Phone: (31) 4116-72496
Fax: (31) 4116-72496

Geological Survey of the Netherlands (G)
Richard Holkane 10, P.O. Box 157
2000 AD Haarlem
Director: F. Bovée
Phone: (31-23)5300300
Fax: (31-23)5351614
Telex: 71105 GEOLD
E-Mail: info@rgd.nl

International Association of Hydrogeologists
Provincial Waterboard of Gelderland
Markstraat 1, Postbus 9090
NL-6800 GX Arnhem
E. Romijn

International Institute of Aerospace Survey and Earth Sciences
Department of Earth Resources Surveys
Box 6
7500 AA Enschede
Phone: (31) 53-320-330
Fax: (31) 53-304-596

International Society of Soil Science
c/o ISRIC
9 Duivendaal
Box 353
NL-6700 AJ Wageningen
J.H.V. Van Baren
Phone: (31)317 471716
Fax: (31)317 4717000
E-Mail: isss@isric.nl

Land Registry Service (C)
Waltersingel 1
Postbus 9046
7300 GH Apeldoorn
Director: J.W.J. Besemer
Phone: (31-55)5285111
Fax: (31-55)5285235

Ministry of Transport and Public Works (H)

Directorate for Waterways
Johan de Wittlaan 3
Postbus 20906
2500 EX x'Gravenhage
Director: R. Adams
Phone: (31-70)3519009
Telex: 31043 HDW NL
E-Mail: radams@hdw.rws.minvenw.nl

Royal Geological and Mining Society of The Netherlands

Koninklijk Nederlands Geologisch
  Mijnbouwkundig Genootschap
Postbus 157
2000 AD Haarlem
L.J. Witte
Phone: (31-23) 5300300
Fax: (31-23) 5367064
http://www.rgd.nl

Rÿks Universiteit Utrecht

Facultÿt Aardwetenschappen (Geoscience)
Postbus 80021
Utrecht 3508 TA

Technische Universiteit Delft

Facuty of Mining and Petroleum Engineering
Postbus 5028
Delft 2600 GA

TNO Institute of Applied Geoscience (G,H)

P.O. Box 6012
2600 JA Delft
Director: H. Speelman F. Walter
Phone: (31)15 2 697184
Telex: 38071 ZP TNO NL
http://www.tno.nl/instit/gg/gg.html

Universiteit van Amsterdam

Facultejt Aardwetenschappen (Geoscience)
Postbus 7161
Amsterdam 1007 MC

NETHERLANDS ANTILLES

Dienst van het Kadaster (Office of the Land Registry) (C)

President Romulo Betancourt Boulevard 4
Willemstad, Curacao
Director: Ir. H.B. Calvo
Phone: (599-9)61118

NEW CALEDONIA

Institut Français de Recherche Scientifique por le Développement en coopération (ORSTOM) (E,G,H)

BP A5
98848 Nouméa Cédex
Director: François Jarrige
Phone: (687) 26 10 00
Fax: (687) 26 43 26
E-Mail: jarrige@noumea.orstom.nc

Service des Mines et de l'Energie (G,R)

1 ter rue E. Unger
B.P. 465
Noumea
Phone: (687) 27 39 44
Fax: (687) 27 23 45

Service Topographique (C)

Territorial Administration Center
Avenue Paul Doumer
Noumea

NEW ZEALAND

University of Auckland

Department of Geology
Private Bag 92019
Auckland
Phone: 64-9-375-7599
Fax: 64-9-373-7435

University of Canterbury

Department of Geology
Private Bag 4800
Christchurch
Phone: 64-3-667-001
Fax: 64-3-642-769
E-Mail: geol015@canterbury.c.nz

Coal Research Limited

Box 31-244
Lower Hutt
R.S. Whitney
Phone: (64-4) 570-3700
Fax: (64-4) 570-3701
E-Mail: r.whitney@crl.co.nz

Geological Society of New Zealand Inc.

Geology Department
University of Otago
P.O. Box 56
Dunedin
Richard Norris

Phone: (64-9) 479 7519
Fax: (64-9) 479 7527
http://www.otago.ac.nz/geology

Institute of Geological and Nuclear Sciences Limited (E,G,H)

Box 30-368
Lower Hutt
Group Manager: John Varcoe
Phone: 64-4-570-1444
Fax: 64-4-569-0600
E-Mail: j.varcoe@gns.cri.nz

Massey University

Department of Soil Science
Private Bag
Palmerston North
Phone: 64-6-356-9099
Fax: 64-6-350-5632

New Zealand Antarctic Programme

Box 14-091
Christchurch
Gillian Wratt
Phone: (64-3) 358 0200
Fax: (64-3) 358 0211
E-Mail: d.shepard@nzap.iac.org.nz

New Zealand Hydrological Society

Box 12-300
Wellington North
Phone: (64 3) 325 6701
Fax: (64 3) 325 2418
E-Mail: faheyb@landcare.cri.nz

University of Otago

Department of Geology
P.O. Box 56
Dunedin
Phone: 64-3-479-7519
Fax: 64-3-479-7527
E-Mail: ougeology@otago.ac.nz

Survey and Land Information Department (C)

P.O. Box 170
Wellington
Surveyor General: W.A. Robertson
Phone: (64-04)4735022
Fax: (64-04)4958450

Victoria University of Wellington

Department of Geology
P.O. Box 600
Wellington
Phone: 64-4-715-345
Fax: 64-4-955-186
E-Mail: geolsec@matai.vuw.ac.nz

University of Waikato

Department of Earth Sciences

Private Bag 3105
Hamilton 2001
Phone: 64-7-838-4024
Fax: 64-7-856-0115
E-Mail: ethsec3@waikato.ac.nz

NICARAGUA

Compania Nacional Productora de Cemento (CANAC) (G)

Apartado 75
Managua
Secretary: Mr. Oscar M. Cerna
Phone: (505-2)662443/661027/660798
Fax: (505-2)662325
Telex: 1037

Corporacion Nicaraguense de Minas (INMINE) (R)

Apartado 195
Managua
Phone: (505-2)52071/52072
Fax: (505-2)51043/50154

Division de Recursos Geotermicos Del Ine (G)

Casa Nazareth 1-1/2c. al Lago
Managua
Director: Ing. Ernesto Martinez Tiffer

Geophysical Observatory

Observatorio Geofisico
Apartado Postal 1761
Managua

Institute of Seismology

Instituto de Investigaciones Sismicas
Apartado 1761
Managua
Roger L. Argenal
Phone: (505-2) 496-987
Fax: (505-2) 496-987

Instituto Nicaraguense de Estudios Territoriales (G)

Complejo Civico Camilo Ortega Saavedra
Managua
Director: Dr. Alejandro Rodriquez

Instituto Nicaraguense de Minas e Hidrocarburos (INMINE) (G)

Centro Comercial El Punto
Managua, D.N.
Ministro: Ing. Ramiro Bermudez Mallol

73

Instituto Nicaraguense de Recursos Naturales y del Ambiente (IRENA) (G)
Apartado Postal No. 5123
Managua, D.N.
Director: Sr. Julio Castillo Ortiz

Ministerio de Construccion y Transporte Instituto Nicaraguese de Estudios Territoriales (INETER) (G)
Apartado 2110
Managua
Director General: Claudio Gutierrez Huete
Phone: (505-2)44739/41890
Fax: (505-2)41890
Telex: 1084 INETER M.

NIGER

Bureau de Recherches Geologiques et Minieres (G)
B.P. 11458
Niamey

Direction de la Recherche Géologique et Minière (G)
Ministere des Mines, Industrie et de la Technologie
B.P. 11700
Niamey
Secretary General: Abdoul Razack Amadou
Phone: (227) 73 45 82
Director of Mines: Issoufou Mahamane

Direction du Service Topographique et du Cadastre (C)
Ministere des Finances
B.P. 250
Niamey
Directeur: Abdou Kane

Service Hydrologique de Genie Rural (H)
Ministere du Developement Rural
B.P. 241
Niamey
Directeur: Ali Seyni

NIGERIA

Faculty of Sciences (E,G)
University of Niamey
P.O. Box: 10.662
Lagos

Dean of Faculty: Professor Alzouma
Phone: (234-1)733072
Fax: (234-1)733862
Telex: 5258 UNINIM

Geological and Mining Research Office (G)
P.O. Box 11458
Lagos
Director: H. De Hays
Phone: (234-1)722325

Geological Survey Department of Nigeria
Ministry of Mines and Power
P.M.B. 2007
Kaduna South
Kaduna State

Ministry of Hydrology and Environment (E,H)

  Directorate of Environment
  P.O. Box 578
  Lagos
  Director: Goumandakoye Mounkaila
  Phone: (234-1)733329
  Fax: (234-1)733329
  Telex: 5509

  Directorate of Water Resources
  P.O. Box 257
  Lagos
  Director: Issa Soumana
  Phone: (234-1)723889
  Fax: (234-1)734642
  Telex: 5509 MIHENVIR

Ministry of Mines, Energy, Industry and Handicraft (C,E,R)

  Directorate of Energy
  Director: Mamadou Zanguina
  Phone: (234-1)734582
  Telex: 5418 NI

  Directorate of Geology and Mining Research
  Director: Hamadou Oumarou
  Phone: (234-1)734582
  Telex: 5418 NI

  Directorate of Mines
  P.O. Box 11.700
  Lagos
  Director: Mallah Hamidou
  Phone: (234-1)734582
  Telex: 5418 NI

National Office of Mining Research (G)
P.O. Box 12.716
Lagos
General Director: Ousmane Gaourdi

Phone: (234-1)735924
Fax: (234-1)732812
Telex: 5300

NORWAY

Geological Survey of Norway (G)
Box 3006 Lade
N-7002 Trondheim
Managing Director: Arne Bjorlykke
Phone: (47-73)904011
Fax: (47-73)921620
Telex: 55417
http://www.ngu.no

The Geophysical Commission (G)
Meteorologisk Institutt
Postboks 43 Blindern
0313 Oslo 3
President: Arne Grammeltvedt
Phone: (47-2)605090
Telex: 21564

International Union for Quaternary Research
Agricultural University of Norway
Department of Soil and Water Sciences
P.O. Box 5028
N-1432 Aas
Sylvi Haldorsen
Phone: (47) 64948252
Fax: (47) 64947485
http://inqua.nlh.no

International Union of Geological Sciences (IUGS) (G)
IUGS Secretariat
Geological Survey of Norway
P.O. Box 3006
N-7002 Trondheim
Executive Secretary: Hanne Refsdal
Phone: (47-73) 90 40 11
Fax: (47-73) 92 16 20
Telex: 55417 NGUN

Norwegian Geological Society
Norsk Geologisk Forening
Postboks 3006 Lade
7002 Trondheim
Arne Solli
Phone: (47-73) 904011
Fax: (47-73) 921620
E-Mail: arne.solli@ngu.no

Norwegian Institute for Water Research (H)
Postboks 33, Blindern
0313 Oslo 3
Chief: Haakon Thaalow
Phone: (47-2)235280
Telex: 74190 NIVAN

Norwegian Petroleum Society
Teknostallen, Prof. Brochsgt. 6
N-7030 Trondheim
Rita OOsterlie MoOrch
Phone: (47-73) 540325
Fax: (47-73) 945510

Norwegian Polar Institute (C,G,E)
Middelthuns Gate 29
P.O. Box 5072, Majorstua
N-0301 Oslo
Director: Olav Orheim
Phone: (47-22)95 95 00
Fax: (47-22)95 95 01
http://www.npolar.no/

Norwegian Water Resources and Energy Administration (H)
Hydrology Department
P.O. Box 5091, Majorstua
0301 Oslo 3
Director: Arne Tollan
Phone: (47-2)469800
Telex: 7939

OMAN

Ministry of Agriculture and Fisheries (H)
P.O. Box 467
Muscat
Minister: Abdul Hafez Salim Rajab
Phone: (968)702066
Telex: 3503 ON

Ministry of Petroleum and Minerals (G)
P.O. Box 551
Muscat
Minister: Said Ahmad Al-Shanfari
Phone: (968)603333
Telex: 3280 PLANDE ON

Public Authority for Water Resources (H)
P.O. Box 5225
Ruwi
Chairman: Abdul Hafiz Salim Rajab
Phone: (968)704188
Telex: 3629 ON

PAKISTAN

Fuel Research Centre - PCSIR (G)
Off University Road
Karachi
Director: Nisar Ahmad
Phone: (92-21)4969806
Fax: (92-21)4966671/4969761

Geological Survey of Pakistan (GSP) (G)
Sariab Road
P.O. Box 15
Quetta
Director General: Dr. Anwaruddin
Phone: (92-81)40020

Geoscience Laboratory
Geological Survey of Pakistan
P.O. Box 1461
Shahzad Town
Islamabad
S.H. Gauhar
Phone: (92-51) 240423
Fax: (92-51) 240223
E-Mail: pd@geolab.sdnpk.undp.org

Hydrocarbon Development Institute of Pakistan (HDIP) (G)
230 Nazimuddin Road F-7/4
P.O. Box 1308
Islamabad
Director General: Hilal A. Raza
Phone: (92-51)920-3958
Fax: (92-51)920-4902
Telex: 5516 HDIP PK
E-Mail: dg@hdipl.isb.erum.com.pk

National Institute of Oceanography (NIO) (G)
37-K, Block 6, P.E.C.H.S.
Karachi 29
Director General: G.S. Quraishee
Phone: (92-21)440460
Telex: 237111 TEC H PAK

Oil and Gas Development Corporation (OGDC) (G)
Building C-6, Masood Mansion
Al-Markaz F-8
Islamabad
Chairman: Zakaudinj Malik
Phone: (92-51)853974
Telex: 5692 OGDC PK

Pakistan Meteorological Department (G)
P.O. Box 8454
Karachi 32

Director General: F.M. Qasim Malik

Pakistan Mineral Development Corporation (PMDC) (R)
13-H/9
Islamabad
Chairman: Ijaz Hussain Malik
Phone: (92-51)250931512201
Fax: (92-51)855374
Telex: 54064 PMDC PK

Pakistan Space and Upper Atmosphere Research Commission (SUPARCO) (C)
P.O. Box 3209
Karachi 75730
Director: S.U. Huk
Phone: (92-21)223031
Telex: 24000 SPACE PK

Resource Development Corporation (RDC) (G)
9th Floor, Sheikh Sultan Trust Building
Beaumont Road
Karachi-4
Phone: (92-21)516206

Sarhad Development Authority
Mineral Testing Laboratory, Plot No. 164-C
Industrial Estate Jamrud Road
Peshawar
S. Hamidullah
Phone: (51) 48675

Survey of Pakistan (C)
P.O. Box 1068
Murree Road
Rawalpindi
Surveyor General of Pakistan
Phone: (92-51)842229
Fax: (92-51)453960

Water and Power Development Authority (WAPDA) (H)
WAPDA House
Lahore
Chairman: Lt. Gen. Zahid Ali Akbar Khan
Phone: (92-42)32223/212722
Telex: 44869 WAPDA PK

PANAMA

Direccion General de Hidrocarburos (G)
Apartado Postal 9658, Zona 4
Director: Hugo Tovar
Phone: (507)227-2035

Direccion General de Recursos Minerales (G)
Ministerio de Comercio e Indistria (MICI)
Apartado Postal 8515
Panama 5
Director: Francia de Sierra
Phone: (507)236-3173
Fax: (507)236-3173

Instituto de Recursos Hidráulicos y Electrificación (H)
Apartado 5285
Panama5
Director: Fernando Aramburú Porras
Phone: (507) 227-2240
Fax: (507) 262-9294

Instituto Geografico Nacional "Tommy Guardia" (C)
Ministerio de Obras Publicas
Apartado Postal 5267, Zona 5
National Director: Jose A.Tejada S.
Phone: (507)236-1843, 236-2534
Fax: (507)236-1841

Instituto Nacional de Acueductos y Alcantarillados (IDAAN) (H)
Apartado 5234, Zona 5
Director: Nilson Espino
Phone: (507)223-7519
Fax: (507)229-3298

Instituto Nacional de Recursos Nacionales Renovables (H)
Apartado Postal 2016 Paraiso
Corregimiento de Ancon
Director: Rolando Guillen
Phone: (507)232-6643
Fax: (507)232-6612

Para Cobre (G)
Apartado Postal 9658, Zona 4
Vice President: Mathew Edler
Phone: (507)223-9945, 223-8945
Fax: (507)263-2449

PAPUA NEW GUINEA

Geological Survey (G)
Office of Minerals and Energy
Box 778
Port Moresby
Chief Geologist: Robin Moaina
Phone: (675)212422
Fax: (675)211360
Telex: NE 23305

University of Papua New Guinea
Geology Department
Box 414
University PO NCD
Phone: 675-267-395
Fax: 675-267-187

The Papua New Guinea University of Technology
Department of Mining Engineering
Private Mail Bag
Lae
Phone: 675-43-434671
Fax: 675-42-4067

Rabaul Volcano Observatory
Box 386
Rabaul
Phone: (675) 921699
Fax: (675) 921004

PARAGUAY

Ciudad Universitaria (C,G,H)
Department of Science and Technology
Km 10, San Lorenzo
Asuncion
Director: Narciso Gonzales
Phone: (595-21)501517

Institute of Basic Sciences
Director: Ing. Luis Alberto Meyer
Phone: (595-21)31274

Military Geodetic Service (C)
Avenida Artigas No. 920 c/ Via Ferrea
Asuncion
Director: Vicente Wenceslao Ortiz
Phone: (595-21)211-139
Fax: (595-21)211-139

Office of Mineral Resources (G)
Ministerio de Obras Publicas y Comunicaciones
Calle Alberdi y Oliva
Asuncion
Director: Ing. Luis Alberto Ruotti
Phone: (595-21)443768
Telex: 162 PY MOPC

PERU

Empresa Minera del Centro del Peru (CENTROMIN PERU) (G)
Apartado 2412

Lima
Presidente: Oscar Posados Perales
Phone: (51-14)367014
Telex: 21231 PE CENTROMIN

Empresa Minera del Peru (MINPERO PERU) (G)

Bernado Monteagudo 222
Orrantia, Lima
Presidente: Manjuel Lescano
Phone: (51-14)620740
Fax: (51-14)638417
Telex: 20122 PE

General Bureau of Aerophotography (C)

Base Aerea "Las Palmas"
Barranco, Lima
Director: Mayor Gral. F.A.P. Ernesto Lindley M.
Phone: (51-14)670538
Telex: 21501 PE

Geological Society of Peru

Sociedad Geologica del Peru
Apartado 2559
Lima 100
Phone: (51-14) 623947

Geophysical Institute of Peru (G)

Casilla 3747
Lima
President: Ronald Woodman
Phone: (51-14) 368437
Fax: (51-14) 370258
E-Mail: jomach@igpcam.pe

Geophysical Society of Peru

Sociedad Geofisica del Peru
Arnaldo Marquez 2277
Lima 11
Jesus Maria

Instituto Geologico Minero y Metalugico (Geological Mining and Metallurgical Institute) INGEMMET (C,G)

Av. Canada1470 - San Borja
Apartado 889
Lima 41
Technical Director: Hugo Rivera Mantilla
Phone: (51-14)224962
Fax: (51-14)768817

National Geographic Institute (C)

Av. Andrés Aramburu1198
Apartado 2038
Lima - 34
Director General: Guillermo Rebagliati Escala

Phone: (51-14)75-3085
Fax: (51-14)75-9810
E-Mail: postmaster@ignperu.gob.pe

National Institute of Natural Resources (INRENA) (C,E,G,H,R)

Calle Diecisiete No. 355, Urb. El Palomar
San Isidro
Lima
Chief: Miguel Ventura Napa
Phone: (51-1)224-3037
Fax: (51-1)224-3218

National Mining Society

Sociedad Nacional de Mineria
Plaza San Martin 917
Lima

National Service of Meteorology and Hydrology (H,R)

Servicio Nacional de Meteorologia e Hidrologia
Jr. Cahuide No. 805, Oficina 1308
Lima 11
Director: Jose Manuel Ames Ruiz
Phone: (51-14)472-4180
Fax: (51-14)471-7287
E-Mail: postmaster@senamh.gob.pe

Naval Bureau of Hydrography and Navigation (C,H)

Calle Saenz Pena No. 590, La Punta
Callao Lima
Director: Luis Moreno Gonzales
Phone: (51-14)652995
E-Mail: hidronav+@amauta.rcp.ne.pe

Peruvian Institute of the Sea (IMARPE) (G,H)

Esq. Gamarra y Javier Valle
Chucuito, Callao
Presidente: Vice Almirante A.P. Ricardo Zevallos
Phone: (51-14)293931

Peruvian Mining and Petroleum Society

Sociedad Nacional de Mineria y Petrolero
Av. Las Flores 346
Lima-27
Roque Benavides
Phone: (51-14) 70-4260
Fax: (51-14) 70-4245

PHILIPPINES

Coast and Geodetic Survey Department (C)
421 Barraca Street
Binondo 1006
Director: Commodore Renato B. Feir
Phone: (63-2)475645/484679

Department of Energy (G)
OEA Building
Philippine National Petroleum Center (PNPC) Complex
Merritt Rd.
1201 Fort Bonifacio
Makati, Metro Manila
Executive Director: Delfin Lazaro
Phone: (63-2)851021
Fax: (63-2)8178603
Telex: 22666 EDC PH

Environmental Management Bureau (E)
Department of Environment and Natural Resources
Topaz Building
99-101 Kamias Road
Quezon City
Director: Manuel S. Gaspay
Phone: (63-2)975698
Fax: (63-2)975698

Geological Society of the Philippines
c/o Mines and Geosciences Bureau
North Avenue, Diliman
Quezon City
Graciano P. Yumull, Jr.
Phone: (63) 920-21-83
Fax: (63) 928-85-44

Mines and Geosciences Bureau (G)
  Department of Environment and Natural Resources
  North Avenue, Diliman
  Quezon City 1104
  Director: Horacio C. Ramos
  Phone: (63-2)928-86-42
  Fax: (63-2)920-16-35
  Telex: 27973 ENVINAR PH

  Geological Survey Division
  Chief: Edwin G. Domingo
  Phone: (63-2)998642

National Institute of Geological Sciences (G)
University of the Philippines
1101 U.P. Campus
Diliman, Quezon City
Head: Graciano P. Yumul, Jr.
Phone: (63-2)9205301
Fax: (63-2)9205301

National Mapping and Resource Information Authority (NAMRIA) (C)
NAMRIA Building
1201 Fort Bonifacio
Makati City, Metro Manila
Administrator: Jose Solis
Phone: (63-2)810-5468, 810-5471, 819-0250
Fax: (63-2)810-5468, 810-2891
E-Mail: namria@sun1.dost.gov.ph

National Water Resources Board (H)
8th Floor NIA Building
Epifanio Delos Santos Avenue, Diliman
Quezon City 1100
Executive Director: Luis M. Sosa
Phone: (63-2)952641
Fax: (63-2)952641

Philippine Atmospheric, Geophysical, and Astronomical Services Administration (G)
Asian Trust Bank Building
1424 Quezon Avenue
Manila 1100
Director: R.L. Kintanar
Phone: (63-2)9228401/8406/8313
Fax: (63-2)9229291
Telex: 42021 PAGASA PM

Philippine Institute of Volcanology and Seismology (PHILVOLCS) (G)
Department of Science and Technology
5 & 6 Floor, Hizon Building
29 Quezon Avenue, Quezon City 1100
Director: Raymundo S. Punongbayan
Phone: (63-2)7126110
Fax: (63-2)7124656
E-Mail: phivolcs@x5.phivolcs.dost.gov.ph

Philippine Society of Mining, Metallurgical, and Geological Engineers
Box 1595
Manila

POLAND

Bureau of Geological Concessions (R)
Ministry of Environmental Protection, Natural Resources and Forestry
52/54 Wawelska St.
00-922 Warsaw

Jacek Wròblewski
Phone: 48 22 251503
Fax: 48 22 251503

Carpathian Balkan Geological Association
Danuta Poprawa, Instytut Geologiczny
PL-31-560 Carakow
Skrzatow 1
Oddzial Karpacki

Central Mining Institute (G)
Plac Gwarkow I
40-951 Katowice
Director: Dr. Jozef Pazdziora

Central Office of Geology
Centralny Urzad Geologii
ul. Jasna 6
PL-00-013 Warsaw

Coordinative and Administrative Body (C)
ul. Jasna 2/4
00-013 Warszawa
President: Prof. Zdzislaw Adamczewski

Geological Society of Poland
Polskie Towarzystwo Geologiczne
ul. Oleandry 2a
30-063 Krakow
Andrzej Slaczka
Phone: (48-12) 33-20-41
Fax: (48-12) 33-22-70
E-Mail: slaczka@ing.uj.edu.pl

Institute of Geological Sciences (G)
Polish Academy of Sciences
Al. Zwirki i Wigury 93
PL-02-089 Warszawa
Director: Andrzej Pszczólkowski
Phone: (48-22)221065
Telex: (48-22)221065
E-Mail: ingpan@asp.biogeo.uw.edu.pl

Institute of Geophysics (G)
ul. Ksiecia Janusza
01-452 Warszawa
Prof. Jerzy Jankowski
Phone: (48-22)364440
Fax: (48-22)370522
http://www.igf.edu.pl

Institute of Meteorology and Water Management (H)
ul. Podlesna 61
01-673 Warszawa
Director: Jan Zielinski
Phone: (48-22)34-18-51
Fax: (48-22)34-54-66

Institute of Paleobiology
Polish Academy of Sciences
PL-02-089
Warsaw, Zwirki i Wigury 93
Phone: (48-2)658 38 19
Fax: (48-2)22 16 52
E-Mail: paleo@asp.biogeo.uw.edu.pl

Instytut Geodezji i Kartografii (C)
ul. Jasna 2/4
00-013 Warszawa
Director: Prof. Bogdan Ney

Instytut Geografii (G)
Przestrzennego
Zagospodarowania PAN
ul. Krakowskie Przedmiescie 30
00-325 Warszawa
Director: Prof. Jerzy Kastrowicki

Instytut Geologiczny (C,E,G,H)
ul. Rakowiecka 4
00-975 Warszawa
Director: St. Speczik
Phone: (48-22)495351
Fax: (48-22)495342
Telex: 815541 IGOL PL
E-Mail: speczik@pgi.waw.pl
http://www.pgi.waw.pl

Muzeum Ziemi PAN (G)
Al. Na Skarpie 20/26
00-488 Warszawa
Director: Dr. Krzysztof Jakubowski

Oil and Gas Institute (G)
ul. Lubicz 25a
31-503 Krakow
Director: Jozef Raczkowski
Phone: (48-12)210033
Fax: (48-12)210050
E-Mail: zndietri@cyf-kr.edu.pl

Zaklad Oceanologii PAN (G)
ul. Powstancow Warszawy 55
81-967 Sopot
Director: Prof. Czeslaw Druet

PORTUGAL

Center of Geology, Institute of Tropical Scientific Research
Centro de Geologia do Instituto de
Investigacao Cientifica Tropical
Al. D. Afonso Henriques 41-4-Dt
P-1000 Lisboa
Ricardo Quadrado
Phone: (351-1) 8476405

Portugal

Direccao-Geral dos Recursos Naturais (H)
Avenida Almirante Gago Coutinho, 30
1000 Lisboa
Director General: Dr. Rui Roda
Phone: (351-1)806094/808001
Telex: 65787 DGRN P

Geological and Mining Institute (C,G,H,R)
Ministério da Economia
R. Almirante Barroso, 38
1050 Lisboa
President: Luis Rodrigues da Costa
Phone: (352-1) 353 75 96
Fax: (351-1) 353 77 09

Geological Society of Portugal
Sociedade Geologica de Portugal
Faculty of Sciences
Lisbon University, R. Escola Politecnica, 58
P-1294 Lisbon Codex

Instituto Português de Cartografia e Cadastro (C)
Rua Artilharia Um, 107
1170 Lisboa
President: Jose Manuel dos Santos Mota
Phone: (351-1)3819600
Fax: (351-1)3819699
http://www/ipcc.pt

Laboratorio Nacional de Engenharia Civil (G,H)
Ministerio de Planeamento e Administracao do Territorio
Avenida Brasil, 101
1799 Lisboa Codex
Director: Prof. Eduardo Romano Arantes e Oliveira
Phone: (351-1)8482131/8482130
Telex: 16760 P

National Institute of Meteorology and Geophysics
Instituto Nacional de Meteorologia e Geofisica
Rua C Do Aeroporto de Lisboa
1700 Lisbon
Tomaz R. Espirito Santo
Phone: 8472880
Fax: 802370

Office of Mines and Geology (G)
Ministerio da Industria e Energia
Rua Antonio Enes 7
1000 Lisboa
Director: Dr. Alcides Rodrigues Pereira
Phone: (351-1)549109/571172
Telex: 62195 GEOMIN P

Portuguese Commission on International Hydrologic Program (H)
Av. Almirante Gago Coutinhao, 30
1000 Lisboa
President: Sr. Rui Roda
Phone: (351-1)808001/809611
Telex: 15853 CHEIAS P

Servicos Geologicos de Portugal (G)
Rua da Academia de Ciencias, 19-2
1200 Lisboa
Director: Dr. Miguel Ramalho
Phone: (351-1)3463915/3460262
Fax: (351-1)3424609

QATAR

Department of Industrial Development (C,G,H)
Ministry of Energy and Industry
P.O. Box 2599
Doha
Director: Sheikh Abdulla Bin Ammed al Thani
Phone: (974)832121
Fax: (974)832024
Telex: 4323 DH

Department of Petroleum Affairs (C,H)
Ministry of Finance and Petroleum
P.O. Box 83
Doha
Director: Abdullah Sallat
Phone: (974)461444
Telex: 4233 QATFIN DH

Water Resources Section (H)
Department of Agriculture and Water Research
P.O. Box 1967
Doha
Chief: Abdel Rahman Al Mahmoud
Phone: (974)417662
Fax: (974)410526
Telex: 4751 HYDAGR DH

REUNION

Observatoir Volcanologique du Piton de la Fournaise
14 R.N.3, 27eme
97418 La Plaine de Cafres

Services des Travaux Publics (G)
St. Denis

ROMANIA

Comisia Romania pentru Activitati Spatiale [Remote Sensing] (G)

Piata Victoriei, 1
Bucharest
President: Constantin Teodorescu
Phone: (40-0)166850, Ext. 1040

Institutul de Geografie (C,E,H)

Str. Dimitrie Racovita2
70307, Bucharest 20
Director: Ion Zavoianu
Phone: (00401)613 59 90
Fax: (00401)311 12 42
E-Mail: janstan@pcnet.pcnet.ro

Institutul de Meteorologie si Hidrologie (H)

Sos, Bucuresti-Ploiesti No. 97
Bucharest
Director: M. Ioana
Phone: (40-1)3129842
Fax: (40-1)3129843
E-Mail: relatumeteo.inmh.ro

Institutul Geologic al Romaniei (G)

Str. Caransebes no. 1
78344 Bucharest-32
Director: Gheorghe Udubasa
Phone: (40-1)665 67 20
Fax: (40-1)312 84 44
E-Mail: udubasa@igr.ro

National Institute for Earth Physics (G)

P.O. Box MG-2
RO-76900 Bucharest-Magurele
Director: Dumitru Enescu
Phone: (40-1)789 76 20
Fax: (40-1)789 76 20
E-Mail: seismos@roifa.ifa.ro

Society of Geological Sciences

Societatea de Stiinte Geologice
Bucharest
Str. Berzei 46

RUSSIAN FEDERATION

All-Russia Petroleum Scientific Research Geological Exploration (VNIGRI)

Liteiny, 39
St. Petersburg 191104
Michael D. Belonin
Phone: (7-812) 273-43-83
Fax: (7-812) 273-73-87
Telex: AT 821208

All-Russian Geological Petroleum Exploration Research Institute (VNIGNI) (G)

36 shosse Entuziastov
Moscow 105118
Director: Constantin A. Kleshev
Phone: (7-095)2732651
Fax: (7-095)2735538
Telex: 207348

All-Russian Research Institute for Geology and Mineral Resources (G)

1, Angliysky Avenue
St. Petersburg 190121
Director: Igor Gramberg
Phone: (7-812)113 83 79
Fax: (7-812)114 14 70
E-Mail: vnilo@g-ocean.spb.su

All-Russian Research Institute for Hydrology & Engineering Geology (VSEGINGEO) (G,H)

142452 Zeleny Village
Noginsk District
Moscow Region
Director: Genrich S. Vartanyan
Phone: (7-095)5212000
Fax: (7-095)9135126
E-Mail: gvartany@sovam.com

All-Russian Scientific Research Institute of Economics of Mineral Resources and Mineral Lands Use (VIEMS) (G)

38, 3-ya Magistralnaya
Moscow 123853
Dirctor: Mikhail A. Komarov
Phone: (095)259 6988
Fax: (095)259 9125
Telex: 113430 GABBRO

All-Union Geological Research Institute

Vasilevsky ostrov
Sredny pr. 74
St. Petersburg

Ammosov Yakutsk State University
58 Belinsky Street
67700 Yakutsk, Sakha Republic
President: Vasily V. Filippov

AMUR Integrated Research Institute (G)
Relochniy, 1
Blagoveshchensk 675000
Director: Valentin G. Moiseyenko
Phone: (8-416-22)42 72 32
Fax: (8-416-22)42 59 31
E-Mail: root@intpmr.amur.su

Bashkortostan State University
32 Frunze Street
450074 Ufa, Bashkortostan
President: Ragib N. Gimaev

Central Research Institute of Geological Prospecting for Base & Precious Metals (TSNIGRI) (E,G)
129B, Varshavskoye Sh.
113545 Moscow
Director: I.F. Migachev
Phone: (7-095)3131818
Fax: (7-095)3152701
E-Mail: geolmos@tsnigri.msk.ru

Chernyshevsky Saratov State University
83 Astrakhanskya Street
410071 Saratov
President: Anatoly M. Bogomolov

Chita Polytechnical Institute
20 Alek.-Zavodskay Street
672039 Chita
President: Yuri N. Reznik

Commission on Loess
International Union for Quaternary Research
Geological Institute
Russian Academy of Sciences
Pyzhevsky per.
Moscow 109017
A.E. Dodonov
Fax: (7-095)231-04-43
E-Mail: dodovov@ginran.msk.su

Committee on Solid Earth Sciences
Pacific Science Association
c/o Russian National Pacific Committee, Russian Academy of Sciences
51 Ulyanovskaya St.
Moscow 109004
N.A. Shilo

Far East Geological Institute (DVGI) (G)
Prospect 100-Letniya 159
Vladivostok 22
Director: Alexander Khenchuk
Phone: (7-4232)318323
E-Mail: fegi@visenet.iasnet.com

Far East Institute of Raw materials (G)
31 Gerasimov St.
Khabarovsk 680005
Director: Yuri I. Barulin
Phone: (7)340659

Far East State Technical University
10 Pushkinkay Street
690600 Vladivostok
President: Gennady P. Turmov

Friendship of Peoples Russian University
6 Miklukho-Maklay Street
117198 Moscow
President: Vladimir M. Filippov

Geological Institute of the Russian Academy of Sciences (GIN RAS) (G)
Pyzhevsky Lane 7
109017 Moscow
Director: Yuri Georgievitch Leonov
Phone: (7-095)231-0443
Fax: (7-095)231-0443
E-Mail: postmast@ginran.msk.su

Geological Research Institute (VSEGEI) (G)
Sredniy Prospekt 74
St. Petersburg 199026
Director: A.D. Shcheglor
Phone: (7-812)2135738
Fax: (7-812)2134418

Gubkin State Academy of Oil and Gas
65 Leninsky Prospect
117917 Moscow V-296, GSP-1
President: Albert I. Vladimirov

Institute for Dynamics of the Geosphere (G)
38 Leninsky Prospect
Moscow 117334
Director: Vitaly Adushkin
Phone: (7-095)9397591
Fax: (7-095)1376511
Telex: 111062
E-Mail: idg@glas.apc.org

Institute of Experimental Mineralogy (IEM) (G)
Moscow District

142432 P/O Chernogolovka
Director: Vilen Andreevich Zharikov
Phone: (7)5245039
http://www.iem.ac.ru

Institute of Geology and Geochemistry (G)

7 Pochtoviy Pereulok
Ekaterinburg 620219
Director: Victor A. Koroteev
Phone: (7)511997

Institute of Geology and Geophysics (IGG) (G)

Universitetsky Prosp. 3
630090 Novosibirsk 90
Director: Nikolai Leontevich Dobretsov
Phone: (7-3952)354661

Institute of Geology of Ore Deposits, Petrography, Mineralogy, Geochemistry (IGEM) (G)

Staromonetny Per. 35
109017 Moscow ZH-17
Director: Nikolai P. Laverov
Phone: (7-095)2314579
Fax: (7-095)2302179

Institute of Lithosphere (ILSAN) (G)

Staromonetny Per. 22
109180 Moscow, ZH-180
Director: N.A. Bogdanov
Phone: (7-095)2335588
Fax: (7-095)2335590
E-Mail: ludmila@ilsan.msk.ru

Institute of Marine Geology and Geophysics (IMGIG) (G)

Ul. Nauki 5
693002 Yuzhno-Sakhalinsk
Director: Konstantin Fedorovich Sergeev
Phone: (7)22128
Telex: 213129

Institute of Physics of the Earth (IFZ) (G)

B. Gruzinskaya, 10
123810 Moscow D-242
Director: Vladimir Nikolaevich Strakhov

Institute of Tectonics and Geophysics (ITIG) (G)

Ul Kim U. Chena 65
680063 Khabarovsk 63
Director: Cherman B. Borukaev
Phone: (7)334205

Institute of the Earth's Crust (IZK) (E,G)

Str. Lermontova 128

664033 Irkutsk 33
Director: Nikolai A. Logachev
Phone: (7-3952)46-40-00
Fax: (7-095)4202106
E-Mail: log@crust.irkutsk.su

Institute of Volcanology (IV) (G)

Ave. Piipa 9
683006 Petropavlovsk-Kamchatka
Director: Boris Ivanov
Phone: (41-500)59195

International Institute of Earthquake Prediction Theory and Mathematical Geophysics (G)

Varshavskoye sh. 79, kor. 2
Moscow 113556
Director: V.I. Keilis-Borok
Phone: (7-095)1107795
Fax: (7-095)3107032
Telex: 411628
E-Mail: mitpan@mitp.rssi.su

Irkutsk State Technical University

83 Lermontov Street
664074 Irkutsk
President: Sergei B. Leonov

Irkutsk State University

1 Karl Marx Street
664003 Irkutsk
President: Fedor K. Schmidt

Kazan State University

18 Lenin Street
420008 Kazan
Republic of Tatarstan
President: Yuri G. Konoplev

Khabarovsk State Technical University

136 Tichookeanskay Street
680035 Khabarovsk
President: Victor K. Bulgakov

Krasnoyarsk Institute of Non-ferrous Metals

95 Prospect Krasnoyarsky Rabochiy
660025 Krasnoyarsk
President: Valery V. Kravstov

Kuban State University

149 K. Libknekht Street
350640 Krasnodar
President: Vladimir A. Babeshko

Limnological Institute (G)

P.O. Box 4199
664033 Irkutsk
Director: Mikhail A. Grachev

Phone: (7-3952)460504
Fax: (7-095)4202106
Telex: 133163

Lomonosov Moscow State University
Department of Geology
Vorobyevy Gory
129805 Moscow
Chair: Boris A. Sokolov
Phone: (095) 939-4220
Fax: (095) 938-0165

Main Computer Center (GLAVNIVTZ) (G)
Russian State Committee on Geology
32A Tuchachevsky St.
Moscow 123585
Phone: (7-095)1928021
Fax: (7-095)1929698
E-Mail: postmaster@glavnivc.msk.su

Moscow Geological Prospecting Institute
23a, Miklucho-Maklaya Street
Moscow 117584

Moscow State Academy of Exploration Geology
23 Miklukho-Maklay Street
117873 Moscow V-485, GSP-7
President: Leonid G. Grabchak

Moscow State Open University
22 P. Korchagin Street
129805 Moscow
President: Anatoly N. Kovshov

National Research Institute of Geology of Foreign Countries (VZG) (E,G)
69, Novocheryomushinskaya Street
117418 Moscow
Director: Evgeny N. Isaev
Phone: (7-095)3325427
Fax: (7-095)4202005
Telex: 412371 VZG

North Caucasus Mining and Metallurgical Institute
44 Nikolaev Street
362021 Vladikavkaz, North Ossetia
President: Zurab M. Khadonov

Northeastern Interdisciplinary Scientific Research Institute (G)
16 Portovaya St.
Magadan 685000
Director: Vladislav I. Goncharov
Phone: (7)30611
Fax: (7)30051
E-Mail: root@neisri.magadan.su

Novocherkassk State Technical University
132 Prosveshenya Street
346400 Novocherkassk, Rostov Region
President: Vitaly A. Taranushich

Novosibirsk State University
2 Pirogova Street
630090 Novosibirsk
President: Vladimir N. Vragov

Okhta Industrial Institute
13 Pervomyskay Street
169400 Okhta, Republic of Komi
President: Gennady V. Rassokhin

Pacific Oceanological Institute, Far Eastern Branch, Russian Academy of Sciences (POI FEBRAS) (G,H,C,E)
43 Baltiyskaya Street
Vladivostok 690041
Director: Viktor Anatolyevich Akulichev
Phone: 7 (423-2)31-1400
Fax: 7 (423-2)31-2573
E-Mail: poi@stv.iasnet.com

Paleontological Society of Russian Academy of Sciences
Sredny 74
St. Petersburg 199026
N. G. Krymgolts
Phone: (7-812) 218-91-56
Fax: (7-812) 213-57-38
E-Mail: vsg@sovam.com

Perm State University
15 Bukireva Street
614600 Perm
President: Vladimir V. Malanin

Plechanov St. Petersburg State Mining Institute
Vasilevsky Ostrov 21 Lane, 2
199026 St. Petersburg
President: Vladimir S. Litvinenko

Rostov State University
105 B. Sadovay Street
199034 Rostov-na-Donu GSP-711
President: Alexander V. Belokon

Russian Academy of Sciences
Geology, Geophysics, Geochemistry and Mining Science Department
Leninsky Prospect 32-a
117334 Moscow
V. A. Zharikov
Phone: (7-095) 938-09-40
Fax: (7-095) 938-19-28
E-Mail: olga@geo.comcp.msk.su

Russian Federation

Russian Federal Service for Geodesy and Cartography (ROSGEOCART) (C)
14 Krzhizanskovo St., Bldg. 2
Moscow 117801
President: Nikolai D. Zhdanov
Phone: (7-095)1243535
Fax: (7-095)1243535
Telex: 411222

Russian State Committee on Geology (ROSKOMNEDRA) (G,R)
B. Gruzinskaya, 4/6
123242 Moscow
Director: Victor P. Orlov
Phone: (7-095)2547633
Fax: (7-095)9430013
Telex: 411772

Scientific Industrial Enterprise on Super Deep Drilling (NEDRA) (G)
Svoboda Str. 8/38
150000 Yaroslavl
Director: B.N. Khakhaev
Phone: (0852)222301
Fax: (0852)328471
E-Mail: postmaster@nedra.yaroslavl.su

Shirshov Institute of Oceanology (IOAN) (G,H)
Ul. Krasikova 23
117218 GSP-7 MOSCOW V-218
Phone: (7-095)1246149
Telex: 411968

St. Petersburg University
Faculty of Geology
Universitetskaya nab., 7-9
St. Petersburg 199034
Dean: Vigor V. Buldakov
Phone: (7-812) 2184418

Stavropol Polytechnical University
2 Prospect Kulakova
355038 Stavropol
President: Boris M. Sinelnikov

Tomsk State University
36 Prospect Lenina
634050 Tomsk
President: Mikhail K. Sviridov

Tver Polytechnical Institute
22 Naberezhnay Afanasya Nikitina
170026 Tver
President: Vycheslav A. Mironov

Tyumen State Technical Oil and Gas University
38 Volodarsky Street
625000 Tyumen
President: Nikolai N. Karnaukhov

Ufa State Oil Technical University
1 Kosmonavtov Street
450062 Ufa, Republic of Bashkortostan
President: Alexander I. Spivakov

Urals State Academy of Mining Geology
30 Kuibyshev Street
620219 Ekaterinburg
President: Ivan V. Dementev

Vernadsky Institute of Geochemistry and Analytical Geochemistry (GEOKHI) (C,E,G)
Ul. Kosygina 19
117975 GSP-1 Moscow V-334
Phone: (7-095)13-4127
Fax: (7-095)938-2054
E-Mail: galimov@geokhi.msk.su

Vinogradov Institute of Geochemistry (IGKH) (G)
Ul. Favorskovo 1A A/YA 701
664033 Irkutsk
Director: Mikhail I. Kuzmin
Phone: (7-3952)460500
Telex: 133163
E-Mail: root@igc.irkutsk.su

Voronezh State University
1 Universitetsky Place
394693 Voronezh
President: Vladimir V. Gusev

RWANDA

Division Cartographie (C)
Ministry of Public Works, Water and Energy
B.P. 24 Kigali
Chief of Division: Minani Camille
Phone: (250)5771

Environment Survey (E)
Ministry of Tourism and Environment
B.P. 2378 Kigali
Director: Rutabingwa Frank
Phone: (250)72095
Fax: (250)76958

Geological Survey of Rwanda (C,G)
Ministry of Trade, Industry and Handicraft
B.P. 73 Kigali
Director of Mines and Geology: Faustin Nyagahima

Phone: (250)73504
Fax: (250)75165

SAMOA

Apia Observatory (E,G,R)
P.O. Box 3020
Apia
Assistant Director: A.K. Titimoec
Phone: (685)20856
Fax: (685)20857

Lands and Survey Department (C)
Main Beach Road
P.O. Box 63
Apia
Director of Lands: Lealiifano Joe Soon
Phone: (685)22481

SAUDI ARABIA

General Petroleum and Mineral Organization (PETROMIN) (G)
P.O. Box 757
Riyadh 11189
Acting Governor: Abdullah Abdul Aziz Al Zaid
Phone: (966-1)4763571
Fax: (966-1)4787141
Telex: 401058 - 401490 SJ

King Fahd University of Petroleum and Minerals (,)

P.O. Box 144
Dhahran International Airport
Box 274, KFUPM
Dhahran
Rector of the University: Dr. Bakr A. Bakr
Phone: (966-3)8602000
Telex: 601060 KFUPM SJ

Department of Earth Sciences
Box 5070
Dhahran 31361
Chairman: Zaki Y. Al-Harari
Phone: (966-3)8602620
Fax: (966-3)8602595
Telex: 601060 KFUPM SJ
E-Mail: es_chair@dpc.kfupm.edu.sa

Ministry of Agriculture and Water (H)

Water Resources Development Department
P.O. Box 478
Riyadh

Director General: Amir Hamad Hussein
Phone: (966-1)4012777/4011699
Telex: 401108 AGRWAT SJ/401692 AGRIRS SJ

Ministry of Petroleum and Mineral Resources (G)
P.O. Box 247
Riyadh
Minister: H.E. Hisham Nazer
Phone: (966-1)4781661/43781133
Telex: 400997/403670 PETROL SJ

Aerial Survey Department
P.O. Box 247
Riyadh
Director General: Soliman Saleh Al-Robaishy
Phone: (966-1)4781661
Telex: 201490/201615/201058

Directorate General of Mineral Resources
P.O. Box 2880
Jiddah
Deputy Minister for Mineral Resources: Ibrahim Al-Khabiri
Phone: (966-2)6671096/6674800
Telex: 401157 DGMR SJ

Saline Water Conservation Corporation (SWCC) (H)
P.O. Box 5968
Riyadh 11432
Governor: Dr. Abdullah Mohammad Al-Ghulaiqa
Phone: (966-1)4630505/4631111/4631763
Telex: 200097/200401/204699 TAHLIA SJ

SENEGAL

Bureau for Cartography and Geology (C,E,H)
7, rue Mohamed V
B.P. 21092-Dakar
Dakar
Director: Aliou Fall
Phone: (221)216226
Fax: (221)213736

Bureau for Geological and Mineral Research (BRGM) (G)
7, rue Mermoz
B.P. 268
Dakar
Director: Didier Fohlen
Phone: (221)227219
Telex: 51274 BRGM-SG

Senegal

International Directory of Geoscience Organizations

Singapore

Bureau of Mines and Geology (G)
Ministry of Industrial Development
Route de Ouakam
B.P. 1238
Dakar
Director: Baidy Diene
Phone: (221)251375
Telex: 61149 MDIA-SG

Commission on Quaternary of Africa
International Union for Quaternary Research, University of Dakar
Department de Geologie, Faculte des Sciences
Dakar
A. Faye

Direction for Urban and Rural Hydraulic (H)
Ministry of Hydraulic and Water Resources
B.P. 2014
Route des Peres Maristes
Dakar
Director: Abdoul-Aziz AW
Phone: (221)324279/325776/323592
Telex: 61302 MHYDRAU SG

Organization for the Development of the Gambia River Basin (OMVG) (H)
BP 2353
Dakar RP
Senegal
Chairman: Malick John
Phone: (221)223159
Fax: (221)225926

ORSTOM Dakar (G)

Route des Peres Maristes-Hann
B.P. 1386
Dakar
Director: Philippo Mathieu
Phone: (221)323480
Fax: (221)324307
E-Mail: mathjieu@dakar.orstom.sn

Geophysical Center
B.P. 50
M'Bour
Director: Ndiath Abdou Saram
Phone: (221)571044
Fax: (221)571500
E-Mail: ndiath@mbour.orstom.sn

Senegal Geographic Service (C)
14, rue Victor Hugo
B.P. 740
Dakar
Director: Serigne M'Baye Thiam
Phone: (221)216567
Telex: 51206 MEINFOM SG

SIERRA LEONE

Geological Survey Division (G)
Ministry of Lands, Mines, and Labor
New England, Freetown
Director: A.E. Agbaye

Surveys and Land Division (G)
Director: E.A. Redwood-Sawyer

Water Resources Division (H)
Ministry of Energy and Power
Leone Building, Siaka Stevens Street
Freetown
Chief Engineer: I.S. Kabia

SINGAPORE

Housing Development Board (C,G)
Building and Development Division
3451 Jalan Bukit Merah
HDB Centre
Singapore 0315
Manager: Yao Chee Liew
Phone: (65)2739090
Fax: (65)2797115
Telex: RS 22020 SINHDB

Jurong Town Corporation (C,G,H)
Technical Division
301 Jurong Town Hall Road
Jurong Town Hall
Singapore 2260
Senior Director: Lim Sak Lan
Phone: (65)5600056
Fax: (65)5655301
Telex: RS 35733 JTC

Ministry of the Environment (C,G,H)
Environmental Engineering Division
40 Scotts Road
Environment Building
Singapore 228231
Director: Chiang Kok Meng
Phone: (65)7327733
Fax: (65)7319456
E-Mail: chiangkm@cs.gov.sg

Public Works Department (G,H)
Ministry of National Development
5 Maxwell Road, -00 and -00
Tower Block, MND Complex
Singapore 069110
Director General: Tan Swan Beng
Phone: (65)2221211
Fax: (65)3258848
http://www.gov.sg/mnd/pwd

Science Centre Board (G,H)
Science Center Building
Science Center Road
Singapore 2260
Scientific Officer: Tey Mui Lee
Phone: (65)5603316
Fax: (65)5659533
Telex: 5603316

Southeast Asia Petroleum Exploration Society (SEAPEX)
Tanglin Box 423
Singapore 9124
Kelley Smoot
Phone: (65) 235 0142
Fax: (65) 735 0013

Survey Department (C)
Ministry of Law
8 Shenton Way -01
Treasury Building
Singapore 068811
Head: Low Oon Song
Phone: (65)2259911
Fax: (65)3239801

SLOVAKIA

Centre of Geoscience Research
Slovak Academy of Sciences
811 06 Bratislava Jozetska 7

Geological Institute of the Slovak Academy of Sciences
Geologicky ustav SAV
Dubravska cesta 9
842 26 Bratislava
Eduard Kohler
Phone: (42-2) 7373961
Fax: (42-2) 7377097

Geological Survey of Slovak Republic (G)
Mlynskadolina 1
817 04 Bratislava
Director: Palol Grecula
Phone: (42-7)373 408
Fax: (42-7)371 940
http://www.guds.sanet.sk

Slovak Hydrometeorological Institute (E,H)
Jeséniova 17
833 15 Bratislava-Koliba
Director: Štefan Škulec
Phone: (42-7)371 247
Fax: (42-7)374 593

Telex: 922 08 HMUBA C
E-Mail: skulec@shmuvax.shmu.sk

SOLOMON ISLANDS

Ministry of Agriculture and Lands (C)
Survey Division
P.O. Box G13
Honiara
Principal Cartographer: Elison Suri
Phone: (677)23567/21511

Ministry of Energy, Water and Mineral Resources (G,H)
Water and Mineral Resources Division
P.O. Box G37
Honiara
Director of Geology: Donn Tolia
Phone: (677)21521
Fax: (677)25811

Ministry of Natural Resources (E)
Environment & Conservation Division
P.O. Box G24
Honiara
Chief Executive Officer: M. Biliki
Phone: (677)21521
Fax: (677)21245
Telex: SOLNAT HQ 66306

SOMALIA

Geological Survey Department (G,H)
Hydrogeological Department
Director: Mohamed Yusuf Awale

Ministry of Minerals and Water Resources
P.O. Box 744
Mogadishu
Director: Mohamed Said Abdi

Hydrogeological Department (H)
Water Development Agency
P.O. Box 525
Mogadishu
General Manager: Khalif Haji Farah

Hydrology Department (H)
Ministry of Agriculture
Mogadishu
Director: Ibrahim Muse

Survey and Mapping Department (C)
Ministry of Defense

P.O. Box 24
Mogadishu
Chief: Colonel Ibrahim Elmi Gelle

SOUTH AFRICA

University of Cape Town
Department of Geochemistry
Rondebosch 7700
Phone: (021) 650-2917
Fax: (021) 650-3783

Department of Geology
Private Bag
Rondebosch 7700
Phone: (021) 650-2931
Fax: (021) 650-3783

Department of Water Affairs (H)
Directorate Geohydrology
Patterson Building
173 Schoeman Street
Pretoria 0002
(Private Bag X313, Pretoria 0001)
Phone: (27-12)299 9111
Fax: (27-12)328 6397
Telex: 322-107
http://www.gov.za/dwaf/index.htme

Directorate, Surveys and Mapping (C)
Private Bag
Mowbray 7700
Chief Director: D.J. Grundlingh
Phone: (27-21)6899721/6854070
Fax: (27-21)6891351
Telex: 521418

Division of Water, Environment and Forestry
Council for Scientific and Industrial Research (CSIR)
P.O. Box 395
Pretoria 0001
Chief Director: A. Yannakoy
Phone: (27-12)8412620
Fax: (27-12)8412689
Telex: 321-312 SA

University of Durban-Westville
Department of Geology
Private Bag X54001
Durban 4000
Phone: (031) 820-2318
Fax: (031) 820-2383

Geological Society of South Africa
Box 44283
2104 Linden

Phone: (27) 888-2288
Fax: (27) 888-1632

Geological Survey (G,H)
280 Pretoria Street
Silverton 0184
(Private Bag X112 Pretoria 00001)
Director: Dr. Cornelius Frick
Phone: (27-12)8411911
Fax: (27-12)8411203
http://www.geoscience.org.za

Magnetic Observatory [Geomagnetism, Seismology] (G)
Council for Scientific and Industrial Research (CSIR)
P.O. Box 32
Hermanus 7200
Director: G.J. Kuhn
Phone: (27-2831)21196
Fax: (27-2831)22039

University of Natal
Dept of Geology & Applied Geology
King George V Avenue
Durban 4001
Phone: (031) 816-2516
Fax: (031) 816-2280

University of Natal (G)
Department of Geology and Applied Geology
Private Bag X10
Dalbridge 4014
Head of Department: F.G. Bell
Phone: (031)260-2516
Fax: (031)260-2280
E-Mail: forbesc@geology.und.ac.za
Head of Geography: Prof. D.H. Davis

University of Natal, Pietermaritzburg
Department of Geology
P.O. Box 375
Pietermaritzburg 3200
Phone: (0331) 955-911
Fax: (0331) 955-599

University of the Orange Free State
Department of Geology
P.O. Box 339
Bloemfontein 9300
Phone: (051) 401-2515
Fax: (051) 474-152

University of the Orange Free State
Institute for Ground Water Studies
P.O. Box 339
Bloemfontein 9300
Director: Prof. F.D.T. Hodgson
Phone: (27-51)4012117

University of Port Elizabeth
Geology Dept
P.O. Box 1600
Port Elizabeth 6000
Phone: (041) 504-2325
Fax: (041) 504-2573

University of Pretoria
Department of Geology
Hillcrest
Pretoria 0083
Phone: (012) 420-2454
Fax: (012) 43-3430

Rand Afrikaans University
Dept of Geology
P.O. Box 524
Johannesburg 2000
Phone: (011) 489-2301
Fax: (011) 489-2309

Rhodes University
P.O. Box 94
Grahamstown, 6140
Head of Geology: Julian S. Marsh
Phone: (27-461)318309
Fax: (27-461)29715
http://www.ru.cic.za/departments/geology

South African Institute of Mining and Metallurgy
13th Floor, Cape Towers
11-13 MacLaren St.
Box 61127
Marshalltown 2107
Phone: (27-11) 834-1273
Fax: (27-11) 838-5923

University of Stellenbach
Department of Geology
Private Bag X5018
Stellenbosch 7600
Phone: (02231) 77-3219
Fax: (02231) 77-4336

University of Stellenbosch
Department of Geology
Private X01, Matieland
Stellenbosch 7602
Phone: (021)808 3219
Fax: (021)808 4336
E-Mail: ar@maties.sun.ac.za

Technikon Pretoria
Department of Materials Technology
Private Bag X 680
Pretoria 0001
Phone: (012)318-6274
Fax: (012)318-6275

University of the Western Cape
Department of Earth Sciences
Geology Section
Private Bag X17
Bellville 7535
Phone: (021) 959-2223
Fax: (021) 951-3627

University of Witwatersrand
Bernard Price Institute for Geophysical Research
Private Bag 3
Wits 2050
Johannesburg
Phone: (011) 716-2430
Fax: (011) 339-7367

Bernard Price Institute for Paleontological Research
Private Bag 3
P.O. Wits 2050
Johannesburg
Phone: (011) 716-2727
Fax: (011) 716-8030

Dept of Geology
Private Bag 3
P.O. Wits 2050
Johannesburg
Phone: (011) 716-2608
Fax: (011) 339-1697

Dept of Geology
Economic Geology Research Unit
Private Bag 3
P.O. Wits 2050
Phone: (011) 716-2742
Fax: (011) 339-1697

Schonland Research Centre
P.O. Wits 2050
Johannesburg
Phone: (011) 716-3166
Fax: (011) 339-2144

University of Zululand
Dept of Geolgy
Private Bag X1001
Kwadlangezwa 3886
Phone: (0351) 9-3911
Fax: (0351) 9-3735

SPAIN

Association of Spanish Geologists
Asociacion de Geologos Espanoles
8-40
28003 Madrid
Reina Victoria

Association of Spanish Petroleum Geologists and Geophysicists
Asociacion de Geologos y Geofisicos Espanoles del Petroleo
Museo Nacional de Ciencias Naturales
Jose Gutierrez Abascal, 2
28006 Madrid
Lorenzo Villalobos
Phone: (341) 348 97 31
Fax: (341) 348 71 20
E-Mail: lvillalobosv@repsol.es

Centro de Estudios Hidrograficos (H)
Paseo Bajo Virgen del Puerto, s/n
28005 Madrid
President: Antonio Nieto Llobet
Phone: (34-1)2656800
Fax: (34-1)26522975
Telex: 22029

Coordinacion de Hidrologica (H)
Direccion General de Obras Hidraulicas
Nuevos Ministerios
28071 Madrid
Chief: Doroteo Frances
Phone: (34-1)2531600, Ext. 2372
Telex: 22285 COCEN E

Universidad de Oviedo (C)
Laboratorio de Geologia
Escuela Superior de Minas
Independencia 13
33004 Oviedo
Director: Jose Antonio Martinez Alvarez
Phone: (34-85)240358
Fax: (34-85)227126
Telex: 84322

Departamento de Geologia (G)
Universidad de Salamanca
Facultad de Ciencias
37008 Salamanca
Director: Jorge Civis Llovera
Phone: (34-23)219763
Fax: (34-23)213619

Direccion General del Instituto Geografico Nacional (C)
General Ibanez Ibero, 3
28003 Madrid
Director General: Angel Arevalo Barroso
Phone: (34-1)2333800
Fax: (34-1)2546743
Telex: 23465 IGC E

Empresa Nacional ADARO de Investigaciones Mineras, S.A. (G)
Doctorsquerdo, 138
28007 Madrid
Director General: Antonio Garcia Moreno

Phone: (34-1)5529900
Fax: (34-1)4335916
Telex: 42083 GEO E

Geological Society of Spain
Sociedad Geológica de España
Fundación Gómez Pardo
Alenza 1
E-28003 Madrid
President: C. Dabrio
Phone: (34-1) 3944817
Fax: (34-1) 3944818

Instituto Andaluz de Ciencias de la Terra (G)
Facultad de Ciencias
18071 Granada
Director: Carlos Sanz de Galdeano
Phone: (34-58)243158
Fax: (34-58)243384
http://dalila.ugr.es/~giact/

Instituto de Ciencias de la Tierra "Jaime Almera" (G)
Lluis Sole i Sabaris, s/n
08028 Barcelona
Director: Angel Lopez
Phone: (34-3)3302716
Fax: (34-3)4110012
http://pangea.ija.csic.es/ija.html

Instituto de Geologica Economica (G)
Facultad de Ciencias Geologicas
Universidad Complutense
28040 Madrid
Director: Alphonso Sopeña
Phone: (34-1)3944813
Fax: (34-1)3944808
E-Mail: sopena@evcmax.sim.ucm.es

Instituto Tecnologico Geominero de Espana (C,G)
Rios Rosas 23
28003 Madrid
General Director: Camilo Caride de Linan
Phone: (34-1)4419245
Fax: (34-1)4426216
Telex: 48054 IGME E

Marine Geology Group
Associated Unity CSIC-UB in Géosciences Marines
Department of Marine Geology and Physical Oceanography
Institute of Marine Sciences
Paseo Joan de Borbo s/n
08039 Barcelona
Department Head: Belén Alonso

Phone: (34-3)2216416
Fax: (34-3)2217340
E-Mail: belen@icm.csic.es

Ministerio de Industria y Energia (G)

Paseo de la Castellana, 160
28046 Madrid
Phone: (34-1)4587058
Fax: (34-1)4578066
Telex: 44204 MISC E

National Commission of Geology

Comision Nacional de Geologia
Rios Rosas 23
ES-28003 Madrid

Real Sociedad Geografica (C)

Valverde 22
Madrid 13

Royal Spanish Society for Natural History

Real Sociedad Espanola de Historia Natural
Facultades de Biología y Geología
Ciudad Universitaria
28040 Madrid
Antonio Perejón
Phone: (34-1) 394 5000
Fax: (34-1) 394 5000

Servicio de Geologia (G)

Ministerio de Obras Publicas y Urbanismo
Avenida de Portugal, 81
28071 Madrid
Director: Bernardo Lopez Camacho
Phone: (34-1)4642371

Spanish Groundwater Club

Alenza, 1
28003 Madrid
Phone: (91) 336 69 78
Fax: (91) 336 69 77

Spanish Society of Paleontology

Sociedad Espanola de Paleontología
Museo Nacional de Ciencias Naturales
José Gutiérrez Abascal, 2
E-28006 Madrid
President: Leandro Sequeiros
Phone: (34-57) 295369
Fax: (34-57) 421864

SRI LANKA

Geological Society of Sri Lanka

Department of Geology
University of Peradeniya
Peradeniya

Phone: (94) 08-88301, EXT 211

Geological Survey and Mines Bureau

No. 4, Senanayake Building
Galle Road
Dehiwala
N.P. Wijayananda
Phone: (94-1) 725745
Fax: (94-1) 735752
E-Mail: gsmb@lka.toolnet.org

Geological Survey Department (G)

48 Sri Jinaratna Road
Colombo 2
Director: D.E.D.S. Jayawardene
Phone: (94-1)24250

Hydrology Section (H)

Irrigation Department
P.O. Box 1138
Colombo 7
Chief Engineer: N.M.G. Navaratne
Phone: (94-1)588879

Surveyor General's Office (C)

Kirula Road
Colombo 5
Surveyor-General: N.C. Seneviratne
Phone: (94-1)585569
Fax: (94-1)584532

Water Resources Board (H)

2A Gregory's Avenue
Bullers Road
Colombo 7
Chairman: K. Yoganathan
Phone: (94-1)696910
Fax: (94-1)696910

SUDAN

Geological and Mineral Resources (G)

Ministry of Energy and Mining
Geological Survey Department
P.O. Box 410
Khartoum
Director: Abdalla Hassan Ishag

Geological Society of Africa

Department of Geology
University of Khartoum
Khartoum

Hydrology Division (H)

Ministry of Agriculture and Irrigation
Nile Waters Department
P.O. Box 878
Khartoum

Director: Isam Mustafa
Ministry of Internal Affairs (C)
Survey Department
P.O. Box 306
Khartoum
General-Director: Dafalla Salih

Rural Water Administration (H)
Ministry of Energy and Mining
P.O. Box 2087
Khartoum
Director-General: Khairalla Mahgoub

SURINAME

Centraal Bureau Luchtkaartering (C)
P.O. Box 971
Maystraat 39
Paramaribo
Head: R.H. Wong Fong Sang
Phone: (597)497246

Geology and Mining Service of Suriname (G)
Ministry of Natural Resources
Klein Waterstraat 2-6
Paramaribo
Director (Acting): G.M. Gemerts
Phone: (597)476215

Hydrologic Research Division (H)
Ministry of Public Works,
   Telecommunication and Architecture
Verlengde Coppenamestraat 167
Paramaribo
Head: Leonard Codrington
Phone: (597)62959

Water Supply Service (H)
Ministry of Natural Resources and Energy
Klein Combeweg 15
Paramaribo
Head: E. Tsai Meu Cheong
Phone: (597)77075

SWAZILAND

Department of Surveys (C)
Ministry of Works and Communications
P.O. Box 58
Mbabane
Surveyor General: A.B. Mhlanga
Phone: (268)42321
Telex: 2301 WD

Geological Survey and Mines Department (G,H,R)
P.O. Box 9
Mbabane
Director: Aaron M. Vilakati
Phone: (268)42411/2/3
Fax: (268)45215
Telex: 2301 WD

Ministry of Natural Resources and Energy (C,E,G,H,R)
P.O. Box 57
Mbabane
Principal Secretary: M.N. Nkambule
Phone: (268)46244/5/6
Fax: (268)42436
Telex: 2301 WD

Rural Water Supply Board (H)
P.O. Box 961
Mbabane
Senior Water Engineer: N.M. Ntezinde
Phone: (268)23231
Telex: 2301 WD

Water Resources Branch (H)
Ministry of Natural Resources and Energy
P.O. Box 57
Mbabane
Senior Water Engineer: Robert Thabede
Phone: (268)42321
Fax: (268)42436
Telex: 2301 WD

SWEDEN

Central Office of the National Land Survey (C)
Lantmaterigatan 2
S-801 82 GAVLE
Director General: Gunilla Olofsson
Phone: (46-26)633000
Fax: (46-26)613277
Telex: 47359 1MV
http://www.lmv.se

Geological Society of Sweden
Geologiska Foreningen i Stockholm
c/o SGU, Box 670
S-751 28 Uppsala
Monica Beckholmen
Phone: (48-18) 358153

Geological Survey of Sweden (C,G,H)
Box 670
S-751 28 Uppsala
Director General: Olof Rydh

Phone: (46-18)179000
Fax: (46-18)179210
http:www.sgu.se

International Geosphere-Biosphere Programme

Royal Swedish Academy of Sciences
Box 50005
S-104 05 Stockholm
Chris Rapley
Phone: (46-8)16 64 48
Fax: (46-8)16 64 05
http://www.igbp.kva.se

Swedish Environmental Protection Agency (E)

Blekholmsterrassen 36
106 48 Stockholm
Director General: Rolf Annerberg
Phone: (46-8)6981000
Fax: (46-8)202925
Telex: 11131 ENVIRON S
http://192.36.242.4

Swedish Meteorological and Hydrological Institute (SMHI) (H,E)

1, Folkborgsvagen
601 76 Norrkoping
Director General: Hans Sandebring
Phone: (46-11)158000
Fax: (46-11)170207
Telex: 64400 SMHI
http://www/smhi.se

The Swedish National Road and Transport Research Institute (G)

Geological Division
S-581 95 Linköping
Director General: Thomas Korsfeldt
Phone: (46-13)20 40 00
Fax: (46-13)14 14 36
E-Mail: hans g johansson@vti.se

National Geotechnical Institute (SGI)
35, Olaus Magnus vag,
S-581 01 Linkoping
Director: Jan Hartlen
Phone: (46-13)115100
Telex: 50125 VTISGI

SWITZERLAND

Federal Division for Water Protection (H)

Monbijoustrasse 43
3003 Bern
President: Marcel Blanc
Phone: (41-31)322 69 69
E-Mail: hans.ulrich.schweizer@buwal.admin.ch

Federal Office for Topographical Survey (C)

Seftigenstrasse 264
CH-3084 Wabern
Director: Francis Jeanrichard
Phone: (41-31)9632111
Fax: (41-31)9632459
E-Mail: francis.jeanrichard@lt.admin.ch

Federal Office for Water Economy (H)

Hallwylstrasse 4
3003 Bern
Director: Dr. Rodolfo Pedroli

Geologisches Institut (G)

Eidgenossische Technische Hochschule
ETH-Zentrum
8092 Zurich
Head: J.-P. Burg
Phone: (41-1)632 6027
http://www.erdw.ethz.ch/

International Association of Sedimentologists

Institut de Géologie
Pérolles
1700 Fribourg
General Secretary: AndreStrasser
Phone: (41)37 29 89 78
Fax: (41)37 29 97 42

International TOGA Project Office

c/o WMO
Box 2300
CH-1211 Geneva 2
J. Marsh
Phone: (41-22) 7308 234/225
Fax: (41-22) 734-3181

International Union for Quaternary Research

Engineering Geology
ETH-Honggerberg
CH-8093 Zurich
Christian Schluchter
Phone: (41-1) 377 25 21
Fax: (41-1) 371 55 48

Landeshydrologie und Geologie (G,H,C,E)

Swiss National Hydrological and Geological Survey
CH-3003 Bern
Director: Ch. Emmeneggen

95

Taiwan Provincial Water Conservancy (H)
Bureau
501 Sec.2, Lee-Min Rd.
Taichung
Director General: Juei-Lin Hsieh
Phone: (886-4)2528841
Fax: (866-4)2529260

Water Resources Planning Commission (H)
Ministry of Economic Affairs
10 F, 41-3, Sec. 3, Hszny Road
Taipei
Chairman: Chian-min Wu
Phone: (866-2)7023473
Fax: (866-2)7542244
E-Mail: wrpcmoea@tpts1.seed.net.tw

TANZANIA, UNITED REPUBLIC OF

University of Dar es Salaam (G)
Ministry of National Education
P.O. Box 35091
Dar es Salaam
Chairman (Geology Dept.): J.T. Nanyaro
Phone: (255-51)49071
Telex: 41561 UNIVIP-TZ

Department of Mineral Resources (G)
P.O. Box 903
Dodoma
Commissioner for Geology: S.L. Bugaisa
Economic Geology: C.M.A. Mosby
Phone: (255-51)22002
Telex: 41777 MAJI TZ

Economic Geology
P.O. Box 903
Dodoma
Semkiwa P.M.
Phone: (255-61)23020/22793
Telex: 53324 MADINI-TZ

Hydrological Section (H)
Ministry of Water
P.O. Box 35066
Dar es Salaam
Senior Principal Engineer: A.K. Muze
Phone: (255-51)48247
Senior Hydrologist: J.M. Kobalyenda and W.S. Lyimo
Principal Geologist: A.M. Mwakinga and G.M. Kifua

Survey and Mapping Division (C)
Mapping Production
Senior Surveyor: O. Andrew
Phone: (255-51)23735

Ministry of Lands, Natural Resources and Tourism
P.O. Box 9201
Dar es Salaam
Commissioner for Survey and Mapping: E.N. Njau
Phone: (255-51)21241

THAILAND

Chiang Mai University
Department of Geological Sciences
Faculty of Science
Chiang Mai 50200
Phone: (66-53)221699
Fax: (66-53)892261

Chulalongkorn University
Department of Geology
Faculty of Science
Bangkok 10330
Phone: (66-2)218-5442
Fax: (66-2)252-9924

Coordinating Committee for Coastal and Offshore Geoscience Programmes in East and Southeast Asia (CCOP)
Offshore Mining Organization Building, 2nd Floor
110/2 Southern Nua Road
Bangrak, Bangkok 10500
Phone: (66-2) 234 3578
Fax: (66-2) 237 1221

Department of Mineral Resources (G)
Economic Geology Division
Director: Boonmai Inthuputi
Phone: (66-2)2456216

Economic Geology Division
Rama VI Road, Rajathevi
Bangkok 10400
Director General: Suvit Sampattavanija
Phone: (66-2)202-3850
Fax: (66-2)202-3702
E-Mail: pipob@mozart.inet.co.th

Environment Division
Director: Prakmard Suwanasing
Phone: (66-2)2479449

Geological Survey Division
Director: Phisit Dheeradilok

Phone: (66-2)202-3735
Fax: (66-2)202-3754

Ground Water Division
Director: Chaeroen Chuamthaisong
Phone: (66-2)2456813

Mineral Fuels Division
Director: Nopadol Mantajit
Phone: (66-2)2478981

Department of Public Works (G,H)

Material and Research Division
Director: Anusornant Mahavinichaimontri
Phone: (66-2)4352700
Fax: (66-2)4331362

Ministry of Interior
218/1 Rama VI Road
Phayathai Bangkok 10400
Director General: Prajaya Sutabutr
Phone: (66-2)2730879
Fax: (66-2)2730879

Water Supply Development Division
Director: Pronchai Kositanurit
Phone: (66-2)2810635
Fax: (66-2)2824880

Geotechnical Engineering International Resources Center

Asian Institute of Technology
P.O. Box 4
Klongluang 12120
Lilia Robles-Austriaco
Phone: (66-2) 524-5862
Fax: (66-2) 516-2126
http://www.ait.ac.th/clair/centers/geirc

Khon Kaen University

Department of Geotechnology
Faculty of Technology
Khon Kaen 40002
Phone: (66-43)24233
Fax: (66-43)239329

Mekong River Commission Secretariat (R)

Kasatsuk Bridge
Rama I Road
Bangkok 10330
Chief Executive Officer: Yasunobu Matoba
Phone: (66-2)2250029
Fax: (66-2)2252796
E-Mail: wolfgang@mozart.inet.co.th

Meteorological Department (E)

Agrometeorological Division
4353 Sukhumvit Road, Phrakanong
Ba Nang, Bangkok 10260
Director: Chalermchai Eg-karntrong

Phone: (66-2)3931682
Fax: (66-2)3939409

Hydrometeorology Division
Meteorological Department
4353 Sukumvit
Bangkok 10260
Director: Kriengkrai Khovadhana
Phone: (66-2)3989868

Studies and Research Division
Director: Patipat Patvivatsiri
Phone: (66-2)3992355

Ministry of Agriculture and Cooperatives (C,G,H)

Ratchadamnoen Nok Road
Bangkok 10200
Permanent Secretary: Sommai Surakul
Phone: (66-2)2810858
Fax: (66-2)2813513

Royal Forestry Department
Director General: Pong Leng-Ee
Phone: (66-2)5791589
Fax: (66-2)5791587

Watershed Management Division
Director: Sawat Dulyapach
Phone: (66-2)5792811
Fax: (66-2)5792811

National Research Council of Thailand (C,G)

Engineering Geology Branch
Chief: Sompit Luadthong
Phone: (66-2)2230021, Ext. 9

Ministry of Science, Technology and Environment
196 Phaholyothin Road
Chatuchak
Bangkok 10900
Secretary General: Suvit Vibulsresth
Phone: (66-2)5612245
Fax: (66-2)5792289
E-Mail: suvit-vibulsresth@fc.nrct.go.th

Remote Sensing Division
Director: Paibul Ruangsiri
Phone: (66-2)5790116
Fax: (66-2)5613035

Surveying and Mapping Branch
Chief: Yongyuth Sitlamuth
Phone: (66-2)2230021, Ext. 229

Petroleum Authority of Thailand (G)

555 Vibhavadi Rangsit Road
Bangkok 10900
Governor: Pala Sookawesh
Phone: (66-2)5373930/1
Fax: (66-2)5373498/9
E-Mail: 220066@ptt.or.th

Royal Irrigation Department (C,G,H)

Geotechnical Division
Director: Chaiwat Preechawit
Phone: (66-2)2413340
Fax: (66-2)5838348

Hydrology Division
Director: Prasert Milintangul
Phone: (66-2)2415217
Fax: (66 2)5838332

Ministry of Agriculture and Cooperatives
811 Samsen Road, Dusit
Bangkok 10200
Director General: Roongrueng Chulajata
Phone: (662)2410065
Fax: (662)2430966

Research and Laboratory Division
Pakkret, Nonthaburi 11120
Director: Vidhaya Samaharn
Phone: (66-2)5838448
Fax: (66-2)5835011

Topographical Survey Division
Director: Narong Sopoak
Phone: (66-2)5838409
Fax: (66-2)2430966

Royal Thai Survey Department (C)
Supreme Command Headquarters
Ministry of Defense
Rachinee Road
Bangkok 10200
Air Chief: Marshall Voranat Aphichari
Phone: (66-2)2806279

TOGO

Departement Hydro-Geologie Ministere des Mines de l'Energie, des Resources Hydrauliques et des Travaux Publics (G,H)
B.P. 356
Lome
Chief: Comlanvi N. D'Almeida

Direction de la Pedologie (G)
Ministere de l'Amenagement Rural
B.P. 1026
Lome
Director: Koffi Lomko Allaglo

Direction Generale des Mines, de la Geologie et du Bureau National de Recherches Minieres (G,H)
B.P. 356
Lome
Director: M. N'zoulou Pere

Ministere de l'Equipement, des Mines et de l'Energie
Direction de la Cartographie Nationale et du Cadastre
B.P. 500
Lome
Director: Raphaël Issa-Gnon Gbarre
Phone: (228) 21 66 15
Fax: (228) 21 68 12

TONGA

Ministry of Lands, Survey and Natural Resources (C,G,H)
P.O. Box 5
Nuku'alofa
Superintendent: S.L. Tongilava

TRINIDAD AND TOBAGO

Geological Society of Trinidad and Tobago
Box 3524
La Romain
Trinidad
Elliston Welsh
Phone: (809) 623-2727
Fax: (809) 623-2726

Institute of Marine Affairs (E,G)
P. O. Box 3160
Carenage P.O.
Director: Lennox Ballah
Phone: (809)6344292
Fax: (809)6344433
E-Mail: director@ima.gov.tt

Ministry of Energy and Energy Industries (G)
Riverside Plaza
P.O. Box 96
Port-of-Spain
Permanent Secretary: Rupert Mends
Phone: (809)6236708/6714
Fax: (809)6232726
Telex: 22714 MENR WG
E-Mail: ttomener@undp.org

Ministry of Planning and Mobilisation (C)
Lands and Surveys Division

Red House
St. Vincent Street
Port-of-Spain
Acting Director of Surveys: Dr. Aldwyn Philip
Phone: (809)6279201/9204
Fax: (809)6245982

Seismic Research Unit (C)
c/o University of the West Indies
St. Augustine
Acting Director: Lloyd Lynch
Phone: (809)6624659
Fax: (809)663-9293
Telex: 24520 UWI WG

TUNISIA

Faculte des Sciences de Tunis
Departement de Geologie
Campus Universitaire
1060 Tunis
Phone: (216-1)512600
Fax: (216-1)885408

General Direction of Studies and Major Hydraulic Works (H)
30 Rue Alain Savary
Tunis
Director General: Adbephofidh Khazen
Phone: (216-1)280034
Telex: 13378 MINAGR

Ministry of Agriculture (E,G,H)
General Direction of Water Resources
43, Rue de la Manoubia
1008 Tunis
Director General: Djemili El Batti
Phone: (216-1)399 320
Fax: (216-1)391 549

Ministry of Energy and Mines (G)
Avenue Mohamed V
Tunis
Director: Bechi Ouni
Phone: (216-1)782668
Telex: 14652

Office National des Mines
BP 215
1080 Tunis Cedex
Director: Abderro P. Touhami
Phone: (216-2)788842
Fax: (216-2)794016

Office of Topography and Cartography (C)
Cite Olympique

Tunis
President Director General: Ali Kallel
Phone: (216-1)782933
Telex: 1412G

TURKEY

Association of Geomorphologists of Turkey
PK 652
Kizilay-Ankara

Chamber of Geological Engineers of Turkey
TMMOB Jeoloji Muhendisleri Odasi
P.K. 464
Kizilay
06424 Ankara
Phone: (90-4) 4343601
Fax: (90-4) 4342388

Chamber of Geophysical Engineers of Turkey
Mithatpasa Cad. 45/15
P.K. 749
Kizilay
06420 Ankara
Mehmet Altintas
Phone: (90-4) 4351379
Fax: (90-4) 4321085
E-Mail: jeofizik@servis9.net.tr

Earthquake Research Department
Deprem Arastirma Dairesi
Box 763
Kizilay-Ankara
Oktay Ergunay
Phone: 4-2873645
Fax: 2137658

General Command of Mapping (C)
Ministry of Defense
T.C., MSB. Harita Genel Komutanligi
06100 Cebeci - Ankara
General Commander: Erdogan Dirik
Phone: (90-312)319 77 40
Fax: (90-312)320 14 95
E-Mail: harita-g@servis2.net.tr

General Directorate of Mineral Research and Exploration (MTA) (C,G,H)
06520 Ankara
Director General: M. Ziya Gozler
Phone: (90-4)2873430
Fax: (90-4)2879188
Telex: 42040 MTA TR

Turkey

General Directorate of Pollution Prevention and Control (E)
Cevre Kirliligi Onleme
Kontrol Genel Mudurlugu
Istanbul Cad. 88 Iskitler
Ankara
General Director: Ali Riza Yilmaz
Phone: (90-4)3423900/3418756
Fax: (90-4)3424001

Mineral Research and Exploration Institute
Maden Tetkik ve Arama Genel Mudurlugu
Eskisehir Yolu Ustu
Ankara
Phone: 4 11791 60
Fax: 4 11815 51

Petrol Ofisi A.S Genel Mudurlugu (G)
Besteker Sok. No. 8
Kavaklidere - Ankara
General Director: Hayrullah Nur Aksu
Phone: 0 0 312 417 64 60
Fax: 0 0 312 425 12 63

State Hydraulic Works (DSI) (H)
Ismet Inönü Bulvari
06100 Yücetepe - Ankara
Director General: Dogan Altinbilek
Phone: (90-312)418 34 09
Fax: (90-312)418 24 98
E-Mail: dogan@knidos.cc.metu.edu.tr

Turkish Association of Petroleum Geologists
Türkiye Petrol Jeologlari Dernegi
Mustafa Kemel Mah.2. Cad No. 86
Esentepe, 06520, Ankara
A. Sami Derman
Phone: (90-312)286 9040
Fax: (90-312)285 5566
E-Mail: derman@cc.tpao.gov.tr

UGANDA

Department of Lands and Surveys (C)
P.O. Box 7061
Kampala
Commissioner: Paul Bakashabaruhanga

Geological Survey and Mines Department (G)
P.O. Box 9
Entebbe
Acting Commissioner: David P.M. Hadoto

Ministry of Natural Resources
Directorate of Water Development
P. O. Box 19
Entebbe
Acting Commissioner: Nsubuga-Senfuma
Phone: (256-42)20914
Fax: (256-42)20971
E-Mail: andjelic@imul.com

UKRAINE

Donetsk Technical University
Dept of Geology
58, Artema Street
Donetsk 340000

Institute of Geologic Sciences (C,E,G,H,R)
Chkalov Street 55b
252054 Kiev
Director: Prof. Peter F. Shpak
Phone: (044)2169496
Fax: (044)219394
Telex: 131444 POISK SU

Ivanovo-Frankivsk Institute for Oil & Gas
Dept of Geology
15, Karpatskaya Street
Ivanovo-Frankivsk 284018

Kharkov State University
Dept of Geology
4, Dzerzhinsky Square
Kharkov 310077

Kiev State University
Dept of Geology
17, Volodimirskaya Street
Kiev 252601

Krivoy Rog Technical University
Dept of Geology
11, XXII Part S'ezd Street
324030

Lviv State University
Dept of Geology
1, University Street
Lviv 290602

Odessa State University
Dept of Geology
2, Petr Veliky Street
Odessa 270057

State University "Lviv Politechnic"
Dept of Geodesy

12, Bandera St.
Lviv-13
290646
Phone: (0322) 724733
Telex: 234139 ORFEJ
E-Mail: ryk%lpi.lviv.ua@ussr.ea.net

UNITED ARAB EMIRATES

Directorate of Intelligence (C)
Topographical Maps, Aerial Maps
Director: Col. Hamad Mahamed Thani
Phone: (971-2)447999
Fax: (971-2)444082

Ministry of Agriculture and Fisheries (H)
Department of Soil and Water
Director: Mahamed Saqar Al-Asam
Phone: (971-4)228161
Fax: (971-4)232781

Ministry of Electricity and Water Desalinization (H)
Phone: (971-2)335099
Fax: (971-2)213738

Ministry of Petroleum and Mineral Resources (C,G)
P.O. Box 59
Abu Dhabi
Director: S.A.S. Al Mahmoudi
Fax: (971-2)663414
Telex: 22544 MPMR EM

Mineral Department
Director: Ahmed Majid
Phone: (971-2)661102
Fax: (971-2)663414
Telex: 22544 MPMR EM

Petroleum Department
Phone: (971-2)668573/651810, Ext. 222

Society of Explorationists in the Emirates
Schlumberger Technical Services
Box 9261
Dubai
Roy Nurmi

UNITED KINGDOM

University of Aberdeen
Department of Geology and Petroleum
 Geology

Merston Building
King's College AB9 2UE
Phone: 0224-273433
Fax: 0224272785
Telex: 73458 UNIABN G

Arthur Holmes Society
Department of Geological Sciences
University of Durham
Science Laboratories
South Road
Durham DH1 3LE
L. Gurney
Phone: (49-91) 374-2520
Fax: (49-91) 374-3740

Birkbeck College
Department of Geology
715 Gresse Street
London W1P 1PA
Phone: 071-631-6508

University of Birmingham
School of Earth Sciences
Edgbaston
Birmingham B15 2TT
Phone: 021-414-3344
Fax: 021-414-3344
Telex: 33762 UOBHAM G

University of Bristol
Department of Geology
Wills Memorial Building
Queens Road
Bristol BS8 1RJ
Phone: 0272-303030
Fax: 0272-253385

British Antarctic Survey (G)
High Cross, Madingley Road
Cambridge CB3 OET
Head, Geoscience Division: M.R.A. Thomson
Phone: (44-123)251444
Fax: (44-123)62616
Telex: 817725 BASCAM G
E-Mail: m.thomson@bas.ac.uk

British Geological Survey (G)
Kingsley Dunham Centre
Keyworth Nottingham NG12 5GG
Director: Peter J. Cook
Phone: (44-115)936 3100
Fax: (44-115)936 3200
Telex: 9378173 BGSKEY G
http://www.nkw.ac.uk/bgs/

British Geotechnical Society
Institution of Civil Engineers
Great George St.
Westminster
London SW1 3AA

A. Guy
Phone: (49-71) 222-7722 EXT. 9829
Fax: (49-71) 222-7500

British Micropalaeontological Society (G)
British Geological Survey
Keyworth
Nottingham NG12 5GG
J.B. Riding
Phone: (49-115) 9363447
Fax: (49-115) 9363437
E-Mail: j.riding@bgs.ac.uk

University of Cambridge
Department of Earth Sciences
Downing Street
Cambridge CB2 3EQ
Phone: 0223-333422
Telex: 811240 CAMSPL G

The Chartered Institution of Water and Environmental Management
15 John St.
London WC1N 2EB
Tony Bispham
Phone: (44-171) 831 3110
Fax: (44-171) 405 4967

Coventry University
School of Natural & Environmental Sciences
Priory Street
Coventry CV1 5FB

Directorate of Military Survey (C)
Ministry of Defence
Elmwood Avenue
Feltham, Middlesex TW13 7AE
Director: Major-General P.E. Fagan MBE
Phone: (44-1)8903622
Telex: 10 890 2148

University of Durham
Department of Geological Sciences
  Laboratories
South Road
Durham
Phone: 091-374-2520
Fax: 091-374-3741
Telex: 537351 DURLIB G

University of East Anglia
School of Environmental Sciences
Norwich NR4 7TJ
Phone: 0603-56161
Fax: 0603-58553
Telex: 975197

East Midlands Geological Society
Rose Cottage
Chapel Lane
Epperstone
Nottingham 146AE
A.J. Filmer
Phone: (44-115) 9663854

University of Edinburgh
Grant Institute of Technology
West Mains Road
Edinburgh EH9 3JW
Phone: 031-650-1000
Fax: 031-668-3184
Telex: 727442 UNIVED G

Edinburgh Geological Society
Grant Institute of Geology
University of Edinburgh
King's Buildings
West Mains Road
Edinburgh EH9 3JW
David H. Land
Phone: (44-31) 650-8536/4843
Fax: (44-31) 668-3184
E-Mail: egeo01@uk.ac.ed.castle

European Association of Geochemistry
Department of Earth Sciences
Downing Street
Cambridge CB2 3EQ
Keith O'Nions
Phone: (44-223) 33-3400

European Association of Science Editors
49 Rossendale Way
London NW1 0XB
Maeve O'Connor
Phone: (44-171) 388-9668
Fax: (44-171) 383-3092
http://www.compulink.co.uk/
~ease-eurscieditors/

European Union of Geosciences
Department of Earth Sciences
University of Cambridge
Downing St.
Cambridge CB2 3EQ
M.I. Johnston
Phone: (44-223) 333421
Fax: (44-223) 333450

University of Exeter
Department of Geology
North Park Road
Exeter EX4 4QE
Phone: 0392-263915
Fax: 0392-263108

The Geological Society
Burlington House

Piccadilly
London W1V 0JU
Richard Bateman
Phone: (44-71) 434-9944
Fax: (44-71) 439-8975

Geologists' Association
Burlington House
Piccadilly
London W1V 9AG
Phone: (44-71) 434-9298

University of Glasgow
Department of Geology
Lilybank Gardens
Glasgow G12 8QQ
Phone: 041-339-8855
Fax: 041-330-4808

Heriot-Watt University
Dept of Petroleum Engineering
Riccarton
Edinburgh EH14 4AS

University of Hull
Department of Geology
Cottingham Road
Hull HU6 7RX
Phone: 0482-465373
Fax: 0482-466340

Hull Geological Society
8 Samman Close
Broadley Ave.
Anlaby
Hull HU10 7HJ
F. Whitham
Phone: (01-482) 650818

Institute of Hydrology (H)
Maclean Building, Crowmarsh Gifford
Wallingford
Oxfordshire OX10 8BB
Director: Prof. B. Williamson
Phone: (44-491)38800
Fax: (44-491)32256
Telex: 849365 HYDROL G

Institution of Mining Engineers
Danum House
South Parade
Doncaster DN1 2DY
W.J.W. Bourne
Phone: (44-302) 320486
Fax: (44-302) 340554

International Association of Geomagnetism and Aeronomy
Physics Unit
Aberdeen University
Aberdeen AB9 2UE
M. Gadsden
Phone: (44) 738 440 358
Fax: (44) 738 440 450

International Association of Hydrogeologists
National Rivers Authority
550 Streetsbrook Road
Solihull
West Midlands B91 1QT
Andrew Skinner
Phone: (44-21) 711 5802
Fax: (44-21) 711 2794

International Glaciological Society
Lensfield Road
Cambridge CB2 1ER
C.S.L. Ommanney
Phone: (44-1223) 355974
Fax: (44-1223) 336543
E-Mail: 100751.1667@compuserve.com

International Union of Crystallography
2 Abbey Square
Chester CH1 2HU
M.H. Dacombe
Phone: (44-1244) 345431
Fax: (44-1244) 344843
http://www.iucr.ac.uk/welcome.html

International Water Supply Association
1 Queen Anne's Gate
London SW1H 9BT
M. J. Slipper
Phone: (44-171)957 4567
Fax: (44-171)222-7243
E-Mail: iwsa@dial.pipex.com

Joint Association of Geophysics
Department of Earth Sciences
Liverpool University
Box 147
Liverpool L69 3BX
J.H. Davies
Phone: (44-151) 794-5254
Fax: (44-151) 794-5170
E-Mail: davies@liv.ac.uk

University of Keele
Department of Geology
Keele
Staffs. ST5 5BG
Phone: 0782-621111
Fax: 0782-715261
Telex: 36113 UNKLIB G

University of Leeds
Department of Earth Sciences
Leeds LS2 9JT

Phone: 0532-431751
Fax: 0532-335259
Telex: 556473 UNILDS G

University of Leicester
Department of Geology
University Road
Leicester LE1 7RH
Phone: 0533-523933
Fax: 0533-522200
Telex: 347250 LEICUN G

University of Liverpool
Department of Earth Sciences
Brownlow Street
Liverpool L69 3BX
Phone: 051-794-5145
Fax: 051-794-5170
Telex: 627095 UNILPL G

Liverpool Geological Society
Liverpool John Moores University
Byrom St.
Liverpool L3 3AF
J.D. Crossley
Phone: (44-51) 231-2120
Fax: (44-51) 298-1014

University of London, Birbeck
Dept of Geology
Birbeck College
Malet Street
London WC1E 7HX
Phone: (071) 380-7333
Fax: (071) 383-0008

University of London, Imperial College
Department of Geology
Imperial College
Prince Consort Road
London SW7 2BP
Phone: 071-589-5111
Fax: 071-584-7596
Telex: 929484 IMPCOL G

University of London, Queen Mary
  Dept of Geography & Earth Science
  Queen Mary College
  Mile End Road
  London E1 4NS
  Phone: (081) 980-4811

  Geomaterials Unit
  Queen Mary & Westfield College
  Mile End Road
  London WC1E 6BT

University of Manchester
Department of Geology
Manchester M13 9PL
Phone: 061-275-3804

Fax: 061-275-3947

Manchester Geological Association
37 Pershore Road
Hollin
Middleton
Manchester M24 6EW
N. Rothwell
Phone: (0161) 6530862

Meteoritical Society
Natural History Museum
Cromwell Road
London SW7 5BD
Monica Grady

Mineralogical Society
41 Queen's Gate
London SW7 5HR
Phone: (44-71) 584-7516
Fax: (44-71) 823-8021
E-Mail: k_murphy@minersoc.demon.co.uk

University of Newcastle Upon Tyne
Department of Geology
Newcastle Upon Tyne
NE1 7RU
Phone: 091-222-6000
Fax: 091-261-1182
Telex: 53654 UNINEW G

The Open University
Department of Earth Sciences
Walton Hall
Milton Keynes MK7 6AA
Phone: 0906-653012
Fax: 0909-653744
Telex: 825061 OU WALT G

Ordnance Survey International (C)
Romsey Road, Maybush
Southhampton SO16 4GU
Director: N. Land
Phone: (44-1703)792139
Fax: (44-1703)792230
Telex: 477 843
E-Mail: nland@ordsvy.govt.uk
http://www.ordsvy.govt.uk/

Oxford University
Department of Earth Sciences
Parks Road
Oxford
Phone: 0865-272-2000

University of Paisley
Dept of Civil Engineering, Geology Section
High Street
Paisley
Renfrewshire PA1 2BE

Palaeontographical Society
British Geological Survey
Keyworth
Nottingham NG12 5GG
S.P. Tunnicliff
Phone: (44-607) 776111

Palaeontological Association
British Antarctic Survey
High Cross
Madingley Road
Cambridge CB3 OET
J.A. Crame
Phone: (44-223) 61188
Fax: (44-223) 62616

Peak District Mines Historical Society
Peak District Mining Museum
Matlock Bath
Derbyshire DE4 3NR
Librarian: R. Paulson
Phone: (44-629) 583834

Petroleum Exploration Society of Great Britain (G)
2nd Floor
17/18 Dover Street
London W1X 3PB
Phone: (0171) 495 6800
Fax: (0171) 495 7800
E-Mail: pesgb@pesgb.domon.co

Photogrammetric Society
Department of Photogrammetry and Surveying
University College London
Gower Street
London WC1E 6BT
Phone: (44-71) 387-7050, EXT 2741
Fax: (44-71) 380-0453
http://web.city.ac.uk/~ag538/photsoc/photsoc.html

University of Reading
Geoscience Teaching Unit
Box 227
Whiteknights
Reading RG6 2AB
Phone: 0734-318940
Fax: 0734-314404
Telex: 847813

Institute for Sedimentology
Box 227
Whiteknights
Reading
Phone: 0734-318941
Fax: 0734-310279
Telex: 847813

Remote Sensing Society
Department of Geography
The University, University Park
Nottingham NG7 2RD
Phone: 44 (0) 1159 515435
Fax: 44 0 1159 515249
http://www.geog.nottingham.ac.uk/rss/

RHB New College
Department of Gelogy
Egham Hill
Egham TW20 OEX
Phone: 0784-434455
Fax: 0784-71780
Telex: 935504

Royal Geographical Society (with The Institute of British Geographers)
1 Kensington Gore
London SW7 2AR
R. Gardner
Phone: (44-0-171) 589-5466
Fax: (44-0-171) 584-4447
http://www.rgs.org

Royal Society
6 Carlton House Terrace
London SW1Y 5AG
P.T. Warren
Phone: (44-71) 839-5561
Fax: (44-71) 930-2170
E-Mail: ezmb013@mailbox.ulcc.ak.uk

Russell Society
116 Gipsy Lane
Kettering
Northants NN16 8UB
P.S. Jackman
Phone: (44-536) 404207
Fax: (44-536) 404272

University of Sheffield
Department of Geology
Beaumont Building
Brookhill
Sheffield S3 7HF
Phone: 0742-768555 EXT. 4945
Fax: 0742-739826
Telex: 54726 UGSHEFG

Sorby Natural History Society
Geological Section
Division of Continuing Education
85 Wilkinson St.
Sheffield SI0 2GJ
Bob Toynton
Phone: (44-742) 768555, EXT 4932

University of Southampton
Department of Geology

Southampton SO9 5NH
Phone: 0703-595000
Fax: 0703-593052

Southampton Oceanography Centre (G)
Empress Dock
European Way
Southampton SO14 3ZH
Director: J. Shepherd
Phone: (44)703-596111
Fax: (44)703-596115
E-Mail: pauline.simpson@soc.sotan.ac.uk

University of St. Andrews
Department of Geology & Geography
Purdie Building
St Andrews, Fife KY16 9ST
Phone: 0334-76161
Fax: 0334-74487
Telex: 76213 SAULIB G

Systematics Association
Department of Geological Sciences
University of Durham
Science Laboratories
South Road
Durham City DH1 3LE
G. Larwood
Phone: (44-91) 374-2501
Fax: (44-91) 374-3741

University College London
Department of Geological Sciences
Gower Street
London
Phone: 071-387-7050
Fax: 071-387-8057
Telex: 28722 UCPHYS G

The University Lancaster
Environmental Science Division
Institute of Environmental
And Biological Sciences
Lancaster LA1 4YQ
Phone: 0524-65201
Fax: 0524-843854
Telex: 65111 LANCUL G

Ussher Society
Tethyan Consultants
Branshaw House
Downgate
Callington
Cornwall PL17 8JX
Colin L. Williams
Phone: (44-579) 370910

University of Wales, Aberystwyth
Institute of Earth Studies
University of Wales
Aberystwyth SY23 3DB
Phone: 0970-62-2606
Telex: 35181ABYUCWG

University of Wales, Bangor
Dept of Ocean Sciences
University of Wales
Bangor

University of Wales, Cardiff
Department of Geology
University of Wales
P.O. Box 914
Cardiff CF1 3YE
Phone: 0222-874830
Fax: 0222-874263
Telex: 498635

Water Data Unit (H)
Department of the Environment
Romney House, 43 Marsham Street
London SW1P 3EB
Head: Derek Edwards
Phone: (44-71)2768238
Fax: (44-71)2768605

World Coal Institute
3 Logan Place
Kensington
London W8 6QN
Phone: (44-71) 373-0799
Fax: (44-71) 835-1408
E-Mail: wci@wcicoal.demon.co.uk

UNITED STATES

Alabama Geological Society
Box 6184
Tuscaloosa, Alabama 35486
Charles C. Smith
Phone: (205) 349-2852

Alaska Division of Geological and Geophysical Surveys
Suite 200
794 University Avenue
Fairbanks, Alaska 99709-3645
Milton A. Wiltse
Phone: (907) 474-7147
Fax: (907) 479-4779

Alaska Geological Society
Box 101288
Anchorage, Alaska 99510
Pete Sloan
Phone: (907) 271-6123

Albuquerque Geological Society
Box 26884
Albuquerque, New Mexico 87125
C.B. Reynolds
Phone: (505) 881-4780

Amarillo Geophysical Society
Box 1431
Amarillo, Texas 79105-1431
Steve N. Thompson
Phone: (806) 378-1026
Fax: (806) 378-8614

American Association for Zoological Nomenclature
National Museum of Natural History
Smithsonian Institution, NHB 163
Washington, D.C. 20560
R.B. Manning
Phone: (202) 357-4668
Fax: (202) 357-3043

American Association of Petroleum Geologists
Box 979
Tulsa, Oklahoma 74101-0979
Lyle F. Baie
Phone: (918) 560-2639
Fax: (918) 560-2626
E-Mail: lbaie@aapg.org

Coast Geological Society
Box 3055
Ventura, California 93006

Eastern Section
West Virginia Geological and Economic Survey
Box 879
Morgantown, West Virginia 26507-0879
Doug Patchen
Phone: (304) 594-233
Fax: (304) 594-2575

Gulf Coast Section
Williamson Oil & Gas Inc.
Box 1772
Shreveport, Louisiana 71166-1772
Robert L. Williamson
Phone: (318) 221-2923

Mid-Continent Section
Enreco Engineering
Box 9838
Amarillo, Texas 79105
Phone: (806) 379-6424

Rocky Mountain Section
224 Clark Avenue
Billings, Montana 59101
Pat Gilbert
Phone: (406) 248-2222

Southwest Section
Swartz Oil Company
4112 College Hills Boulevard, Suite 201
San Angelo, Texas 76904-6591
Bruce R. Swartz

American Association of Stratigraphic Palynologists
Amoco, Exploration and Production Technology Group
Box 3092
Houston, Texas 77253-3092
David T. Pocknall
Phone: (713) 366-5399
Fax: (713) 366-3195
http://www.geology.utoronto.ca/aasp

American Chemical Society/Geochemistry Division
Chemistry Department
Florida Institute of Technology
150 W. University Boulevard
Melbourne, Florida 32901-6988
Mary Sohn
Phone: (407) 768-8000, ext. 7379
Fax: (407) 984-8461
E-Mail: sohn@roo.fit.edu

American Congress on Surveying and Mapping
5410 Grosvenor Lane
Bethesda, Maryland 20814
John Lisack Jr.
Phone: (301) 493-0200
Fax: (301) 493-8245

American Crystallographic Association
Box 96, Ellicott Station
Buffalo, New York 14205-0096
Marcia Vair
Phone: (716) 856-9600, ext. 321
Fax: (716) 852-4846
E-Mail: marcia@awi.buffalo.edu

American Geological Institute
4220 King Street
Alexandria, Virginia 22302-1502
Marcus E. Milling, Executive Director
Phone: (703) 379-2480
Fax: (703) 379-7563
E-Mail: agi@agi.umd.edu
http://www.agiweb.org/

American Geophysical Union
2000 Florida Avenue, N.W.
Washington, D.C. 20009
A.F. Spilhaus Jr., Executive Director
Phone: (202) 462-6900
Fax: (202) 328-0566

E-Mail: cust_ser@kosmos.agu.org
http://www.agu.org

American Ground Water Trust
16 Centre Street
Concord, New Hampshire 03301
Andrew Stone
Phone: (603) 228-5444
Fax: (603) 228-6557

American Institute of Hydrology
2499 Rice Street, Suite 135
St. Paul, Minnesota
Helen Klose
Phone: (612) 484-8169
Fax: (612) 484-8357
E-Mail: aihydro@aol.com

American Institute of Professional Geologists
Suite 103
7828 Vance Drive
Arvada, Colorado 80003
William V. Knight
Phone: (303) 431-0831
Fax: (303) 431-1332

American Lunar Society
Box 209
East Pittsburgh, Pennsylvania 15112
David O. Darling
Phone: (608) 837-6054

American Quaternary Association
Department of Geosciences
University of Massachusetts
Amherst, Massachusetts 01003-5820
Julie Brigham-Grette
Phone: (413) 545-4840
Fax: (413) 545-1200
E-Mail: brigham-grette@geolgeog.umass.edu
http://www.geo.umass.edu

American Society for Photogrammetry and Remote Sensing
Suite 210
5410 Grosvenor Lane
Bethesda, Maryland 20814-2192
William D. French, CAE
Phone: (301) 493-0290
Fax: (301) 493-0208
E-Mail: billf@asprs.org
http://www.asprs.org/asprs

American Water Resources Association
Suite 300
950 Herndon Parkway
Herndon, Virginia 22070
Kenneth D. Reid

Phone: (703) 904-1225
Fax: (703) 904-1228

Appalachian Geological Society
Box 2605
Charleston, West Virginia 25329
Larry Cavallo

Arctic Research Commission
Suite 630
4350 N. Fairfax Drive
Arlington, Virginia 22203
Dr. Garrett W. Brass
Phone: (703) 525-0111
Fax: (703) 525-0114

Arizona Geological Society
Box 40952
Tucson, Arizona 85717
Mark Miller

Arizona Geological Survey
Suite 100
416 W. Congress Street
Tucson, Arizona 85701
Larry D. Fellows
Phone: (520) 770-3500
Fax: (520) 770-3505

University of Arizona Geophysical Society
Department of Geosciences
Gould-Simpson Building
University of Arizona
Tucson, Arizona 85721
Norman M. Meader
Phone: (520) 621-4845
Fax: (520) 621-2672
E-Mail: meader@geo.arizona.edu

Arkansas Geological Commission
Vardelle Parham Geology Center
3815 W. Roosevelt Road
Little Rock, Ark. 72204
William V. Bush
Phone: (501) 296-1877
Fax: (501) 663-7360

Arthur Lakes Library
Colorado School of Mines
1400 Illinois, Box 4029
Golden, Colorado 80401-0029
Joanne V. Lerud
Phone: (303) 273-3911
Fax: (303) 273-3199
E-Mail: jlerud@flint.mines.colorado.edu

ASFE/Professional Firms Practicing in the Geosciences
Suite G106

8811 Colesville Road
Silver Spring, Maryland 20910
John P. Bachner
Phone: (301) 565-2733
Fax: (301) 589-2017

Association for Women Geoscientists

4779 126th Street N
White Bear Lake, Minnesota 55110- 5910
Dr. Jeanette Leete
Phone: (612) 426-3316
Fax: (612) 426-5449
E-Mail: leete@macalstr.edu
http://flint.mines.edu:45021

Association of American State Geologists

Florida Geological Survey
903 W. Tennessee Street
Tallahassee, Florida 32304-7700
Walter Schmidt
Phone: (904) 488-4191
Fax: (904) 488-8086
E-Mail: schmidt_w@dep.state.fl.us

Association of Earth Science Editors

1101 W. 21st Street
Lawrence, Kansas 66046
Arly H. Allen
Phone: (913) 842-1732
Fax: (913) 843-2089

Association of Engineering Geologists

Suite 2D
323 Boston Post Road
Sudbury, Massachusetts 01776
Edwin A. Blackey Jr.
Phone: (508) 443-4639
Fax: (508) 443-2948

Association of Exploration Geochemists

1252 S. Duddley Street
Lakewood, Colorado 80226
R.K. Glanzman

Association of Ground Water Scientists and Engineers

2600 Ground Water Way
Columbus, Ohio 43219
Jacqueline Mack
Phone: (614) 337-1949, (800) 551-7379
Fax: (614) 337-8445

Association of Women Soil Scientists

6 South Townview Lane
Newark, Delaware 19711
Maria Sadushy Pautler

Atlanta Geological Society

Bruce J. O'Connor
19 Martin Luther King Jr. Dr., SW, Room 400
Atlanta, Georgia 30334
c/o Georgia Geologic Survey
Phone: (404) 656-3214
Fax: (404) 657-8379
E-Mail: 76775.2303@compuserve.com
http://www.westga.edu/geology/web_htm/ags.html

Atlantic Coastal Plain Geological Association

Delaware Geological Survey
University of Delaware
Newark, Delaware 19716
Kelvin W. Ramsey
Phone: (302) 831-3586
Fax: (302) 831-3579
E-Mail: 04432@udel.edu

Austin Geological Society

Box 1302
Austin, Texas 78767
Carolyn Condon
Phone: (512) 454-4797

Baton Rouge Geological Society

Box 82215
Baton Rouge, Louisiana 70884-2215
William H. Schramm
Phone: (504) 765-0610
Fax: (504) 765-0602
http://www.tyrell.net/home1/senm/brgs/

Bay Area Geophysical Society

633 Post Street
San Francisco, California 94109
Phone: (415) 842-3789

Baylor Geological Society

Department of Geology
Baylor University
Box 97354
Waco, Texas 76798-7354
Phone: (817) 755-2361
Fax: (817) 755-2673
E-Mail: cleavy@earthlogic.baylor.edu

Big Rivers Area Geological Society

Route 2, Box 647
Mayfield, Kentucky 42066
Lynn Shelby
Phone: (502) 762-6761

Board on Infrastructure and the Constructed Environment

National Academy of Sciences

2101 Constitution Avenue, N.W.
Washington, D.C. 20418
Richard Little, Director
Phone: (202) 334-3376
Fax: (202) 334-3370

Board on Natural Disasters and U.S. National Committee for the Decade for Natural Disaster Reduction
National Research Council, HA-370
2101 Constitution Avenue, N.W.
Washington, D.C. 20418
Caroline L. Clarke
Phone: (202) 334-1964
Fax: (202) 334-1377

California Division of Mines and Geology
801 K Street, MS 12-30
Sacramento, California 95814-3531
James F. Davis
Phone: (916) 445-1923
Fax: (916) 445-5718

California Earthquake Society
Box 1222
San Marcos, California 92079-1222
Steven G. Spear
Phone: (619) 744-1150, ext. 2513

California Groundwater Association
Box 14369
Santa Rosa, California 95402-6369
Mike Mortensson
Phone: (707) 578-4408
Fax: (707) 546-4906
E-Mail: weelguy@a.crl.com
http://www.groundh2o.org

Carolina Geological Society
Box 90234
Department of Geology
Duke University
Durham, North Carolina 27708-0234
Duncan Heron
Phone: (919) 684-5321
Fax: (919) 684-5833
E-Mail: heron@geo.duke.edu
http://www.geo.duke/cgsinfo.htm

Center for the Earth Sciences
Kean College
Union, New Jersey 07083
Paul Rockman
Phone: (908) 527-2894
Fax: (908) 355-5143

The Clay Minerals Society
P.O. Box 4416
Boulder, Colorado 80306

Patricia Jo Eberl, Manager
Phone: (303) 444-6405
Fax: (303) 444-2260
E-Mail: peberl@clays.org
http://ctjrs.agry.purdue.edu/claymin/clayminsoc.htm

Coastal Bend Geophysical Society
Box 2741
Corpus Christi, Texas 78403
John Smythe

Colorado Geological Survey
Department of Natural Resources
Room 715
1313 Sherman Street
Denver, Colorado 80203
Vicki W. Cowart, Director and State Geologist
Phone: (303) 866-2611
Fax: (303) 866-2461

Colorado Water Congress
Suite 312
1390 Logan
Denver, Colorado 80203
Richard D. MacRavey
Phone: (303) 837-0812
Fax: (303) 837-1607

Commission on Glaciation (INQUA)
c/o Illinois State Geological Survey
615 East Peabody Drive
Champaign, IL 61820
A.K. Hansel
Phone: (217) 333-5852
Fax: (217) 333-2830
E-Mail: hansel@geoserv.isgs.uiuc.edu

Committee on Coral Reefs
Pacific Science Association
The Marine Laboratory
University of Guam
UOG Station
Mangilao, Guam 96923
Charles Birkeland

Commodity Futures Trading Commission
Office of Public Affairs
3 Lafayetter Centre
1155 21st Street, N.W.
Washington, D.C. 20581
Phone: (202) 418-5080
Fax: (202) 418-5525

Connecticut Geological and Natural History Survey
79 Elm Street, Store Level
Hartford, Connecticut 06106-5127

Richard Hyde
Phone: (860) 424-3540
Fax: 424-4058

Conservation and Survey Division
University of Nebraska-Lincoln
113 Nebraska Hall
Lincoln, Nebraska 68588-0517
Perry B. Wigley
Phone: (402) 472-3471
Fax: (402) 472-2410.
E-Mail: pwigley@unlinfo.unl.edu

Council on Undergraduate Research (Geology Division)
Department of Geoscience
Hobart and William Smith Colleges
Geneva, New York 14456
Donald L. Woodrow
Phone: (315) 781-3604
Fax: (315) 781-3587
E-Mail: woodrow@hws.edu

Cushman Foundation for Foraminiferal Research
Invertebrate Paleontology
Museum of Comparative Zoology
Harvard University
26 Oxford Street
Cambridge, Massachusetts 02138
Frederick J. Collier
Phone: (617) 496-5406
Fax: (617) 496-5535
E-Mail: collier@mcz.harvard.edu

Dallas Geological Society
One Energy Square, Suite 223
Dallas, Texas 75206
Phone: (214) 373-8614

Dallas Paleontological Society
Box 710265
Dallas, Texas 75371
Rocky Manning
Phone: (214) 698-4033

Dallas Society of Potential Field Geophysicists
PFG Inc.
6522 Barnsbury Court
Dallas, Texas 75248-1405
James S. Russell-Redman
Phone: (214) 250-0118
Fax: (214) 250-0118

Defense Mapping Agency (DMA) (C)
8613 Lee Highway
Fairfax, Virginia 22031-2137
Director: Major General Raymund E. O'Mara

Phone: (703)2859290
Fax: (703)2859374

Delaware Geological Survey
DGS Building
University of Delaware
Newark, Delaware 19716-7501
Robert R. Jordan
Phone: (302) 831-2833
Fax: (302) 831-3579

Denver Geophysical Society
Suite 1250
518 17th Street
Denver, Colorado 80202
Terry Donze
Phone: (303) 573-3846

Dinamation International Society
550 Jurassic Court
Fruita, Colorado 81521
Phone: (800) 344-3466
Fax: (303) 858-3532

Earthquake Engineering Research Institute
Suite 320
499 14th St.
Oakland, California 94612-1902
Susan K. Tubbesing
Phone: (510) 451-0905
Fax: (510) 451-5411
E-Mail: eeri@eeri.org
http://www.eeri.org

East Tennessee Geological Society
Box 6193
Oak Ridge, Tennessee 37831

Economic Geology Publishing Company
Yale University
Box 208110
New Haven, Connecticut 06520-8110
Brian J. Skinner
Phone: (203) 432-3166
Fax: (203) 432-9819
E-Mail: economic.geology@yale.edu

El Paso Geological Society
Department of Geological Sciences
University of Texas
El Paso, Texas 79968
Brian Penn
Phone: (915) 747-5501
Fax: (914) 747-5073

Environmental and Engineering Geophysical Society
10200 N. 44th Avenue, Suite 304

Wheat Ridge, Colorado 80037
Phone: (303) 422-7905
http://www.esd.ornl.gov/eegs

Environmental Research Institute of Michigan
Box 134001
Ann Arbor, Michigan 48113-4001
Joseph Christy
Phone: (313) 994-1200, ext. 3638
Fax: (313) 665-6559

Field Conference of Pennsylvania Geologists
Box 5871
Harrisburg, Pennsylvania 17110-0871
Phone: (717)787-2169
Fax: (717) 783-7267

Florida Geological Survey
903 W. Tennessee Street
Tallahassee, Florida 32304-7700
Walter Schmidt
Phone: (904) 488-4191
Fax: (904) 488-8086

Florida Paleontological Society Inc.
Florida Museum of Natural History
Gainesville, Florida 32611
Phone: (904) 392-1721
Fax: (904) 392-8783

Four Corners Geological Society
Box 1501
Durango, Colorado 81301-1501
Tom Ann Casey
E-Mail: 75342.3463@compuserve.com

Friends of Mineralogy
14403 Carolcrest
Houston, Texas 77079
Albert L. Kidwell
Phone: (713) 497-1066

Colorado Chapter
Box 150410
Lakewood, Colorado 80215-0401
Ed Raines

Pacific Northwest Chapter
Box 11
Seattle University
Seattle, Washington 98122-4460
Robert J. Smith
Phone: (206) 296-5943
Fax: (206) 461-5235

Friends of Sherlock Holmes
3900 Tunlaw Road N.W., 19
Washington, D.C. 20007-4830
Peter E. Blau

Phone: (202) 338-1808
E-Mail: pblau@capaccess.org

Geochemical Society
Department of Terrestrial Magnetism
5241 Broad Branch Road N.W.
Washington, D.C. 20015
S.B. Shirey
Phone: (202) 686-4387
Fax: (202) 364-8726

Geological Association of New Jersey
Division of Coastal Resources
New Jersey Department of Environmental Protection
CN-401
Pal Building
Trenton, New Jersey 08625-0401
Susan D. Halsey

Geological Society
Department of Earth Sciences
East Texas State University
Commerce, Texas 75428
Phone: (903) 886-5445

Geological Society of America
3300 Penrose Place
Box 9140
Boulder, Colorado 80301
Donald M. Davidson Jr.
Phone: (303) 447-2020
Fax: (303) 447-1133
E-Mail: davidson@geosociety.org

Cordilleran Section
Department of Geology
California State University
Fresno, California 93740
Bruce A. Blackerby
Phone: (209) 278-2955
Fax: (209) 278-5980
E-Mail: bruce@csufresno.edu
http://www.geosociety.org/sectdiv/cord/index.htm

North-Central Section
University Iowa Hygenic Laboratory
102 Oakdale Campus
Iowa City, Iowa 52242-5002
George R. Hallberg
Phone: (319) 335-4500
Fax: (319) 335-4600

Northeastern Section
2300 St. Paul Street
Baltimore, Maryland 21218-5210
Kenneth N. Weaver
Phone: (410) 554-5532
Fax: (410) 554-5502

Rocky Mountain Section
Department of Geology and Geological Engineering
Colorado School of Mines
Golden, Colorado 80401
Kenneth E. Kolm
Phone: (303) 273-3932

South-Central Section
Department of Geology
Baylor University
Box 97354
Waco, Texas 76798-7354
Rena M. Bohem
Phone: (817) 755-2361
Fax: (817) 755-2673

Southeastern Section
Department of Geology
Box 870338
University of Alabama
Tuscaloosa, Alabama 35487-0338
Harold Stowell
Phone: (205) 348-5098
Fax: (205) 348-0818
E-Mail: hstowell@wgs.geo.ua.edu
http://www.geo.ua.edu/segsa/segsa.html

Geological Society of Iowa

Geological Society of Iowa
109 Trowbridge Hall
Iowa City, Iowa 52242-1319
Paul VanDorpe
Phone: (319) 335-1580
Fax: (319) 335-2754
E-Mail: pvandorpe@gsbth-po.igsb.uiowa.edu
http://www.igsb.uiowa.edu/htmls/gsi/gsi.htm

Geological Society of Kentucky

Kentucky Geological Survey
228 Mining and Mineral Resources Bldg.
University of Kentucky
Lexington, Kentucky 40506-0107
Kenneth W. Kuehn
Phone: (606) 257-5500
E-Mail: kwente@kgs.mm.uky.edu

Geological Society of Maine

198 Main Street
Yarmouth, Maine 04096
Fred Beck
Phone: (207) 846-9065
Fax: (207) 846-9066
E-Mail: fmbeck@agate.com

Geological Society of the Oregon Country

Box 907, University Station
Portland, Oregon 97207
Clay R. Kelleher

Geological Society of Washington, D.C.

U.S. Geological Survey
954 National Center
Reston, Virginia 20191
Bruce Lipin
Phone: (703) 648-6100
Fax: (703) 648-6383

Geological Survey of Alabama

Box O
420 Hackberry Lane
Tuscaloosa, Alabama 35486-9780
James D. Moore
Phone: (205) 349-2852
Fax: (205) 349-2861
E-Mail: gsa@ogb.gsa.tuscaloosa.alus
http://www.gsa.tuscaloosa.al.us

Geologist of Washington, D.C.

Department of Biology and Environmental Science
MB 44-04
University of the District of Columbia
4200 Connecticut Avenue, N.W.
Washington, D.C. 20008-1154
James V. O'Connor
Phone: (202) 274-5886
Fax: (202) 274-5589

Geophysical Society of Alaska

ARCO Alaska Inc.
Box 100360
Anchorage, Alaska 99510-0360
Michael T. Wiley
Phone: (907) 265-6976
Fax: (907) 265-1657

Geophysical Society of Houston

Epic Geophysical
1221 Lamar 5
Houston, Texas 77010
Wulf Massell
Phone: (713) 650-3820
E-Mail: geowulf@aol.com

Georgia Geologic Survey

Department of Natural Resources
Room 400
19 Martin Luther King Jr. Drive S.W.
Atlanta, Georgia 30334
William H. McLemore
Phone: (404) 656-3214
Fax: (404) 657-8379

Georgia Geological Society

Department of Geology
Georgia State University
University Plaza
Atlanta, Georgia 30303-3083
John D. Costello

Phone: (404) 651-2272
Fax: (404) 651-2013

Geoscience Information Society
c/o American Geological Institute
4220 King Street
Alexandria, Virginia 22302-1502
Lisa Dunn
Phone: (303) 273-3687
Fax: (303) 273-3199
E-Mail: ldunn@mines.colorado.edu

Geothermal Energy Association
Box 598
Davis, California 95617-1350
David N. Anderson
Phone: (916) 758-2360
Fax: (916) 758-2839
E-Mail: geores@wheek.dcn.davis.ca.us

Geothermal Resources Council
Box 1350
Davis, California 95617
David N. Anderson
Phone: (916) 758-2360
Fax: (916) 758-2839
E-Mail: geores@wheel.dcn.davis.ca.us

Global Systems Science Project
Lawrence Hall of Science
University of California
Berkeley, California 94720-5200
Cary Sneider
Phone: (510) 642-0552
Fax: (510) 642-0552

Graham Geological Society
Box 862
Graham, Texas 76450
Glenn H. Felderhoff

Grand Canyon National Park Research Library
Box 129
Grand Canyon, Arizona 86023
Sara Stebbins
Phone: (602) 638-7768

Grand Junction Geological Society
707 Brassie Drive
Grand Junction, Colorado 81506- 3911
William L. Chenoweth
Phone: (970) 242-9062
Fax: (970) 242-0402

Gulf Coast Association of Geological Societies
Box 672
Austin, Texas 78767
Birdena T. Schroeder

Phone: (512) 282-2354
Fax: (512) 282-8150

Hawaii Department of Land and Natural Resources
Division of Water and Land Development
Box 373
Honolulu, Hawaii 96809
Manabu Tagomori
Phone: (808) 587-0230
Fax: (808) 587-0283

History of the Earth Sciences Society
Department of History
Texas Tech University
Lubbock, Texas 79409-1013
Ronald Rainger
Phone: (806) 742-3744
E-Mail: j3ron@ttacs.ttu.edu

Houston Geological Society
7171 Harwin
Houston, Texas 77036
Suite 314
Phone: (713) 785-6402
Fax: (713) 785-0553

Humboldt Earthquake Education Center
Humboldt State University
Arcata, California 95521
Lori Dengler
Phone: (707) 826-6019
Fax: (707) 826-5241
E-Mail: denglerl@axe.humboldt.edu

Idaho Association of Professional Geologists
Box 7584
Boise, Idaho 83707
T.W. Holland
Phone: (208) 336-8631

Idaho Geological Survey
Room 332
Morrill Hall
University of Idaho
Moscow, Idaho 83843
Earl H. Bennett
Phone: (208) 885-7991
Fax: (208) 885-5826
E-Mail: bennett@aspen.csrv.uidaho.edu

Illinois Geological Society
3121 Lime
Mt. Vernon, Illinois 62864
Mike Tucker

Illinois Groundwater Association
2204 Griffith Drive

Champaign, Illinois 61820
Steve Wilson
Phone: (217) 333-0956

Illinois State Geological Survey
Natural Resources Building
615 E. Peabody Drive
Champaign, Illinois 61820-6964
William W. Shilts
Phone: (217) 333-4747
Fax: (217) 244-7004
E-Mail: isgs@geoserv.isgs.uiuc.edu

Indiana Geological Survey
611 N. Walnut Grove
Bloomington, Indiana 47405
Norman C. Hester
Phone: (812) 855-9350
Fax: (812) 855-2862
http://www.indiana.edu/~igs/index.html

Inland Geological Society
San Bernardino County Museum
2024 Orange Tree Lane
Redland, California 92374
Steven Mains
Phone: (909) 780-4170
Fax: (909) 780-3837

Institute for Geophysics
University of Texas at Austin
8701 N. MoPac Expressway
Austin, Texas 78759-8397
Paul L. Stoffa
Phone: (512) 471-6156
Fax: (512) 471-8844
E-Mail: utig@utig.ig.utexas.edu
http://www.ig.utexas.edu

International Association for Great Lakes Research
Business Office
2200 Bonisteel Boulevard
Ann Arbor, Michigan 48109- 2099
Phone: (313) 747-1673
Fax: (313) 747-2748

International Association for Mathematical Geology
West Virginia Geological Survey
Box 879
Morgantown, West Virginia 26507-0879
Michael E. Hohn
Phone: (304) 594-2331
Fax: (304) 594-2575
E-Mail: hohn@geosrv.wvnet.edu

International Association for the Physical Sciences of the Ocean
Box 810 440
Vicksburg, Mississippi 39182-0440
Fred E. Camfield
Phone: (610) 684-2012
Fax: (610) 634-3483
E-Mail: camfield@vicksburg.com

International Association of Geochemistry and Cosmochemistry
Department of Geological Sciences
Ohio State University
275 Mendenhall Laboratory
125 South Oval Mall
Columbus, OH 43210-1398
Gunter Faure
Phone: (614) 292-2721
Fax: (614) 292-7688

International Association of Hydrogeologists/U.S. National Chapter
2614 Checkerberry Court
Reston, Virginia 22091
P. Patrick Leahy
Phone: (703) 648-6287
Fax: (703) 648-4227
E-Mail: pleahy@usgs.gov

International Center for Arid and Semiarid Land Studies
Mail Stop 1036
Box 41036
Texas Tech University
Lubbock, Texas 79409
Idris Rhea Traylor
Phone: (806) 742-2218
Fax: (806) 742-1954

International Federation of Palynological Societies
Bilby Research Center
Northern Arizona University
Flagstaff, Arizona 86001
R. Scott Anderson
Phone: (520) 523-5821
Fax: (520) 523-7290
E-Mail: sa@nauvax.ucc.nau.edu
http://geo.arizona.edu/palynology

International Landslide Research Group
3262 Ross Road
Palo Alto, California 94303
Earl Brabb
Phone: (415) 329-5140
Fax: (415) 329-4936
http://irpi.unipg.it/ilrg/welcome.html

International Marine Minerals Society
811 Olomehani Street
Honolulu, Hawaii 96813-5513

Phone: (808) 522-5611
Fax: (808) 522-5618
E-Mail: 70673.534@compuserve.com

International Permafrost Association
Box 9200
Arlington, Virginia 22219-0200
Jerry Brown
Phone: (703) 525-3136
Fax: (202) 328-0566
E-Mail: jerrybrown@igc.apc.org
http://www.geodata.soton.ac.uk/ipa

International Tsunami Information Center
737 Bishop Street, Suite 2200
Honolulu, Hawaii 96813-3212
Charles S. McCreery
Phone: (808) 532-6422
Fax: (808) 532-5576
E-Mail: itic@ptwc.noaa.gov

International Union for Quaternary Research
Illinois State Geological Survey
615 East Peabody Drive
Champaign, Illinois 61820
A.K. Hansel
Phone: (217) 333-5852
Fax: (217) 333-2830
E-Mail: hansel@geoserv.isgs.uiuc.edu

Iowa Department of Natural Resources
Geological Survey Bureau
109 Trowbridge Hall
Iowa City, Iowa 52242-1319
Donald L. Koch
Phone: (319) 335-1575
Fax: (319) 335-2754
E-Mail: dkoch@gsbth-po.igsb.uiowa.edu
http://www.igsb.uiowa.edu

Iowa Department of Natural Resources
Geological Survey Bureau
123 N. Capitol Street
Iowa City, Iowa 52242
Donald L. Koch
Phone: (319) 335-1575
Fax: (319) 335-2754

IRIS
1616 N. Fort Myer Drive
Suite 1050
Arlington, Virginia 22209
David Simpson
Phone: (703) 524-6222
Fax: (703) 527-7256
E-Mail: simpson@iris.edu
http://www.iris.edu

IRIS Data Management Center
1408 N.E. 45th Street, Suite 201
Seatte, Washington 98105
Timothy Ahern
Phone: (206) 547-0393
Fax: (206) 547-1093
E-Mail: tim@iris.washington.edu

Joint Oceanographic Institute
Suite 800
1755 Massachusetts Avenue, N.W.
Washington, D.C. 20036-2102
James Watkins
Phone: (202) 232-3900
Fax: (202) 232-8203

Kansas Geological Society and Library
212 N. Market, Suite 100
Wichita, Kansas 67202
Tim Dugan
Phone: (316) 265-8676
Fax: (316) 265-1013

Kansas Geological Survey
1930 Constant Avenue
West Campus
University of Kansas
Lawrence, Kansas 66047
Lee C. Gerhard
Phone: (913) 864-3965
Fax: (913) 864-5317
http://www.kgs.ukans.edu

Kentucky Geological Survey
228 Mining and Mineral Resources Building
University of Kentucky
Lexington, Kentucky 40506-0107
Donald C. Haney
Phone: (606) 257-5500
Fax: (606) 257-1147

The Kentucky Paleontological Society, Inc.
365 Cromwell Way
Lexington, Kentucky 40503
Daniel J. Phelps
Phone: (606) 277-3148
E-Mail: chestnut@fido.mm.uky.edu

Lafayette Geological Society
Box 51896 O.C.S.
Lafayette, Louisiana 70505
Tim Rynott

Lamont-Doherty Earth Observatory
Columbia University
Palisades, New York 10964
John Mutter
Phone: (914) 359-2900
Fax: (914) 365-8162

Louisiana Geological Survey
Box G
Baton Rouge, Louisiana 70893
William E. Marsalis
Phone: (504) 388-5320
Fax: (504) 388-5328

Lunar and Planetary Institute
3600 Bay Area Boulevard
Houston, Texas 77058
David C. Black
Phone: (713) 486-2180
Fax: (713) 486-2173

Maine Geological Survey
Robert G. Marvinney
Department of Conservation
22 State House Station
Augusta, Maine 04333-0022
National Resources Information and Mapping Center
Phone: (207) 287-2801
Fax: (207) 287-2353

Maryland Geological Survey
2300 St. Paul Street
Baltimore, Maryland 21218-5210
Emery T. Cleaves, Director
Phone: (410) 554-5559
Fax: (410) 554-5502
E-Mail: ecleaves@mgs.dnr.md.gov
http://mgs/dnr.md.gov

Massachusetts Office of Environmental Affairs
Room 2000, 20th Floor
100 Cambridge Street
Boston, Massachusetts 02202
Richard N. Foster
Phone: (617) 727-9800
Fax: (617) 727-2754

Michigan Basin Geological Society
Department of Geological Sciences
Michigan State University
206 Natural Science Building
East Lansing, Michigan 48824-1115
Tom Hoane
Phone: (616) 676-2090
Fax: (616) 676-0396

Michigan Earth Science Teachers Association
9001 Hackberry Street
Plymouth, Michigan 48170-4110
Ardis Maciolek
Phone: (313) 343-2289
Fax: (313) 343-0980
E-Mail: amacio@www.sceince.wdyne.edu
http://cirrus.sprl.umich.edu/mesta

Michigan Geological Survey Division
Box 30256
Lansing, Michigan 48909-7756
Phone: (517) 334-6923
Fax: (517) 334-6038

Mineralogical Society of America
1015 18th Street, NW, Suite 601
Washington, D.C. 20036-5203
J. Alexander Speer
Phone: (202) 775-4344
Fax: (202) 775-0018
http://geology/.smith.edu/msa/msa.html

Minerals Management Service
United States Department of the Interior
Office of Public Affairs
1849 C Street, N.W.
Washington, D.C. 20240
Barry A. Williamson
Phone: (202) 208-3983
Fax: (202) 208-3918
http://www.mms.gov

Minnesota Geological Survey
University of Minnesota
2642 University Avenue
St. Paul, Minnesota 55114-1057
David Southwick
Phone: (612) 627-4780
Fax: (612) 627-4778
E-Mail: mgs@gold.tc.umn.edu
http://www.geo.umn.edu:80/mgs/

Mississippi Geological Society
Box 422
Jackson, Mississippi 39205- 0422
Steve Ingram
Phone: (601) 961-5534

Mississippi Office of Geology
Box 20307
Jackson, Mississippi 39289-1307
S. Cragin Knox
Phone: (601) 961-5500
Fax: (601) 961-5521

Missouri Department of Natural Resources
Division of Geology and Land Survey
Box 250
111 Fairgrounds Road
Rolla, Missouri 65402
James Hadley Williams
Phone: (314) 368-2100
Fax: (314) 368-2111

Montana Bureau of Mines and Geology
Montana Tech of the University of Montana

1300 W. Park Street
Butte, Montana 59701-8997
John C. Steinmetz
Phone: (406) 496-4180
Fax: (406) 496-4451

Montana Geological Society

Box 844
Billings, Montana 59103
Steven W. VanDelinder
Phone: (406) 322-4146

Monterey Bay Geological Society

Moss Landing Marine Laboratories
Box 450
Moss Landing, California 95039
H. Gary Greene
Phone: (408) 633-3304
Fax: (408) 753-2826
E-Mail: greene@mlml.calstate.edu

National Aeronautics and Space Administration

Two Independence Square
300 E Street, S.W.
Washington, D.C. 20546
Administrator: Daniel S. Goldin
Phone: (202) 358-0000
Fax: (202) 358-0071

National Association of Black Geologists and Geophysicists

Amoco Production Co.
Box 50879
New Orleans, Louisiana 70150-0879
Patricia Hall
Phone: (504) 586-6973
Fax: (504) 586-2637

National Association of Geoscience Teachers

Box 5443
Bellingham, Washington 98227-5443
Robert Christman
Phone: (360) 650-3587
Fax: (360) 650-7302
E-Mail: xman@henson.cc.wwu.edu

National Association of State Boards of Geology

Box 11591
Columbia, South Carolina 29211
Sandra Swinehart
Phone: (803) 799-1047
Fax: (803) 252-3432

National Center for Earthquake Engineering Research

State University of New York
Red Jacket Quadrangle
Buffalo, New York 14261
George C. Lee
Phone: (716) 645-3391
Fax: (716) 645-3399
E-Mail: nceer@ubvm.cc.buffalo.edu

National Earth Orientation Service

U.S. Naval Observatory
Washington, D.C. 20392-5420
Dennis D. McCarthy
Phone: (202) 762-1887
Fax: (202) 762-1563

National Earth Science Teachers Association

2000 Florida Avenue, N.W.
Washington, D.C. 20009
M. Frank Watt Ireton
Phone: (202) 462-6910, ext. 243
Fax: (202) 328-0566
E-Mail: fireton@kosmos.agu.org

National Geodetic Survey

SSMC3, Room 8657, N/NGS
1315 East-West Highway
Silver Spring, Maryland 20910
Lewis A. Lapine
Phone: (301) 713-3222
Fax: (301) 713-4175
http://www.ngs.noaa.gov

National Ground Water Association

2600 Ground Water Way
Columbus, Ohio 43219
Kevin McCray
Phone: (614) 337-1949, (800) 551-7379
Fax: (614) 337-8445
E-Mail: h2o@h2o-ng wa.org
http://www.h2o-ngwa.org

National Ground Water Information Center

6375 Riverside Drive
Dublin, Ohio 43017
Janet Bix
Phone: (614) 761-3222
Fax: (614) 761-3446

National Mining Association

1130 17th Street, N.W.
Washington, D.C. 20036
Richard L. Lawson
Phone: (202) 463-2625
Fax: (202) 463-6152

National Mining Hall of Fame and Museum

120 W. Ninth Street

Box 981
Leadville, Colorado 80461
Carl Miller
Phone: (719) 486-1229
Fax: (719) 486-3927

National Oceanic and Atmospheric Administration (NOAA) (C)

14th Street and Constitution Ave., N.W.
Washington, D.C. 20230
Under Secretary: D. James Baker
Phone: (202) 482-6090
Fax: (202) 482-6203

National Science Foundation

4201 Wilson Boulevard
Arlington, Virginia 22030
Director: Neal F. Lane
Phone: (703)306-1234
Fax: (202)306-0215

Division of Earth Sciences
4201 Wilson Boulevard
Arlington, Virginia 22030
Ian D. MacGregor
Phone: (703) 306-1550
Fax: (703) 306-0382
E-Mail: imacgreg@nsf.gov

National Speleological Society

2813 Cave Avenue
Huntsville, Alabama 35810- 4431
David Luckins
Phone: (205) 852-1300
Fax: 851-9241
E-Mail: nss@caves.org
http://www.caves.org

Nebraska Geological Society

9810 Florence Heights Boulevard
Omaha, Nebraska 68112
Dean Christensen

Nevada Bureau of Mines and Geology

Stop 178
University of Nevada
Reno, Nev. 89557-0088
Jonathan G. Price
Phone: (702) 784-6691
Fax: (702) 784-1709
E-Mail: jprice@nbmg.unr.edu

New England Intercollegiate Geological Conference

Geology Department
Boston University
Boston, Massachusetts 02215
Dabney Caldwell
Phone: (617) 353-2534
Fax: (617) 353-3290

New England States Emergency Consortium (NESEC)

Lakeside Office Park
607 North Avenue, Suite 16
Wakefield, Massachusetts 01880
Edward S. Fratto
Phone: (617) 224-9876
Fax: (617) 224-4350
E-Mail: NESEC@mail.serve.com
http://www.serve.com/NESEC

New Hampshire Geological Survey

Department of Environmental Services
Box 2008
Concord, New Hampshire 03302-2008
Eugene L. Boudette
Phone: (603) 271-3406
Fax: (603) 271-6588

New Jersey Geological Survey

29 Artcic Parkway
CN-427
Trenton, New Jersey 08625
Haig F. Kasabach
Phone: (609) 292-1185
Fax: (609) 633-1004
http:\\www.state.nj.us/dep/njgs/njgs-html

New Mexico Bureau of Mines and Mineral Resources

Campus Station
Socorro, New Mexico 87801
Charles E. Chapin
Phone: (505) 835-5420
Fax: (505) 835-6333
E-Mail: bureau@gis.nmt.edu
http://geoinfo.nmt.edu

New Mexico Bureau of Mines and Mineral Resources

Campus Station
Socorro, New Mexico 87801
Charles E. Chapin
Phone: (505) 835-5420
Fax: (505) 835-6333

New Mexico Geological Society Inc.

801 Leroy Place
Socorro, New Mexico 87801-4796
David A. Schoderbek
Phone: (505) 835-5410
Fax: (505) 835-6333
http://www.nmt.edu:80/~nmbmmr/nmgs

New Orleans Geological Society

234 Loyola Avenue, Suite 932
New Orleans, Louisiana 70112-2016
Phone: (504) 561-8980

New York State Geological Association
New York State Geological Survey
3140 Cultural Education Center
Albany, New York 12230
William M. Kelly
Phone: (518) 474-7559
Fax: (518) 473-8496
E-Mail: wkelly@museum.nysed.gov

New York State Geological Survey
New York State Museum
3140 Cultural Education Center
Empire State Plaza
Albany, New York 12230
Robert H. Fakundiny
Phone: (518) 474-5816
Fax: (518) 473-8496
E-Mail: rfakundi@museum.nysed.gov

North Carolina Geological Survey
Environment, Health and Natural Resources
Box 27687
Raleigh, North Carolina 27611-7687
Charles H. Gardner
Phone: (919) 733-3833
Fax: (919) 733-4407

North Dakota Geological Survey
600 East Boulevard
Bismarck, North Dakota 58505-0840
John P. Bluemle
Phone: (701) 328-9700
Fax: (701) 328-9898

North Texas Geological Society
Box 1671
Wichita Falls, Texas 76307
Bob Carter
Phone: (817) 592-0402

Northeastern Science Foundation Inc., affiliated with Brooklyn College of the City University of New York
Rensselaer Center for Applied Geology
Box 746
Troy, New York 12181-0746
Gerald M. Friedman
Phone: (518) 273-3247
Fax: (518) 273-3249
E-Mail: friedg2@rpi.edu

Northern California Geological Society
9 Bramblewood Court
Danville, California 94506-1130
Dan Day
Phone: (510) 294-7530
Fax: (510) 455-8362

Northern Illinois University Geophysical Society
Department of Geology
Northern Illinois University
DeKalb, Illinois 60115
C. Patrick Ervin
Phone: (815) 753-1942
Fax: (815) 753-1945
E-Mail: ervin@geol.niu.edu

Northern Ohio Geological Society
Department of Geological Sciences
Case Western Reserve University
Cleveland, Ohio 44106-7216
Philip Banks
Phone: (216) 368-3690
Fax: (216) 368-3691
E-Mail: pob@po.cwru.edu

Northwest Geological Society
10 N. Post Street, Suite 414
Shoreline, Washington 98177-0772
Tim S. Olson
Phone: (206) 546-3868

Northwest Mining Association
414 Peyton Building
Spokane, Washington 99201
Karl W. Mote
Phone: (509) 624-1158
Fax: (509) 623-1241

Northwest Petroleum Association
Box 6679
Portland, Oregon 97228
C.J. Newhouse
Phone: (503) 224-2156

Nuclear Regulatory Commission
One White Flint North Building
11555 Rockville Pike
Rockville, Maryland 20852
Phone: (301) 415-8200

Ocean Drilling Program
Texas A&M University Research Park
1000 Discovery Drive
College Station, Texas 77845-9547
Philip D. Rabinowitz
Phone: (409) 845-2673
Fax: (409) 845-1026

Ohio Division of Geological Survey
Department of Natural Resources
4383 Fountain Square Drive
Columbus, Ohio 43224-1362
Thomas M. Berg
Phone: (614) 265-6576
Fax: (614) 447-1918

Ohio Geological Society
Box 14322
Beechwold Station
Columbus, Ohio 43214
Richard McClish

Oklahoma City Geological Society Inc.
227-B Park Avenue
Oklahoma City, Oklahoma 73102-4405
Steven D. Bridges

Oklahoma Geological Survey
Energy Center
Room N-131
100 E. Boyd
Norman, Oklahoma 73019-0628
Charles J. Mankin
Phone: (405) 325-3031
Fax: (405) 325-7069
E-Mail: cjmankin@uoknor.edu
http://www.uoknor.edu/special/ogs-pttc/

Oregon Department of Geology and Mineral Industries
Suite 965
800 N.E. Oregon Street,
Portland, Oregon 97232-2162
Donald A. Hull
Phone: (503) 731-4100
Fax: (503) 731-4066
http://sarvis.dogami.state.or.us

Pacific Northwest National Laboratories
Box 999
MSIN K9-34
Earth Sciences Department
Richland, Washington 99352
Richard L. Skaggs
Phone: (509) 372-6256
Fax: (509) 372-6153
E-Mail: rl_skaggs@pnl.gov

Paleontological Research Institution
1259 Trumansburg Road
Ithaca, New York 14850-1398
Warren D. Allmon
Phone: (607) 273-6623
Fax: (607) 273-6620
E-Mail: wda1@cornell.edu
http;//www.englib.cornell.edu/pri/

Paleontological Society
University of Chicago
Department of Geophysical Sciences
5734 South Ellis Avenue
Chicago, Illinois 60637
J. John Sepkoski, Jr.
Phone: (312) 702-8167
Fax: (312) 702-9505
E-Mail: jjsepkos@midway.uchicago.edu
http://www.uic.edu/orgs/paleo/homepage.html

Pander Society
Department of Geosciences
University of Missouri
Columbia, Missouri 65211
R.L. Ethington
Phone: (314) 882-6470
Fax: (314) 882-5458
E-Mail: geoscray@showme.missouri.edu

Panhandle Geological Society
Box 2473
Amarillo, Texas 79105
Kerry Rice

Pennsylvania Bureau of Topographic and Geologic Survey
Department of Environmental Resources
Box 8453
Harrisburg, Pennsylvania 17105-8453
Donald M. Hoskins
Phone: (717) 787-2169
Fax: (717) 783-7267

Pittsburgh Association of Petroleum Geologists
Box 16352
Pittsburgh, Pennsylvania 15242
Patrick M. Imbraguo
Phone: (412) 299-7058

Pittsburgh Geological Society
c/o Geomega Inc.
Box 3432
Pittsburgh, Pennsylvania 15230
Reginald P. Briggs
Phone: (412) 835-5506
Fax: (412) 835-5506

Puerto Rico Geological Survey Division
Department of Natural Resources
Box 5887
Puerta de Tierra Station
San Juan, Puerto Rico 00906
Lisbeth Hyman
Phone: (809) 722-2526
Fax: (809) 724-0365

Rhode Island State Geologist
Department of Geology
University of Rhode Island
Kingston, Rhode Island 02881
J. Allan Cain
Phone: (401) 874-2265
Fax: (401) 874-2190
E-Mail: jacain@uriacc.uri.edu
http://www.uri.edu/artsci/gel/

Rocky Mountain Association of Geologists
Suite 505
820 16th Street
Denver, Colorado 80202
Dee Tyler
Phone: (303) 573-8621
Fax: (303) 628-0546

Roswell Geological Society Inc.
Box 1171
Roswell, New Mexico 88201
Phelps White IV
Phone: (505) 622-1001

Sacramento Petroleum Association
Box 254443
Sacramento, California 95865
Owen Kittredge
Phone: (916) 638-2085
Fax: (916) 638-8385
E-Mail: delta@qulknet

San Angelo Geological Society
Box 2568
San Angelo, Texas 76902
John Rissler
Phone: (915) 655-0070

San Diego Society of Natural History
San Diego Natural History Museum
Box 1390
San Diego, California 92112
Michael Hager
Phone: (619) 232-3821
Fax: (619) 232-0248

San Diego State University Geology Alumni
Department of Geological Sciences
San Diego State University
San Diego, California 92182
Monte Marshall
Phone: (619) 595-1395
Fax: (619) 594-4372

Seismological Laboratory (MS 174)
University of Nevada
Reno, Nevada 89557-0141
James N. Brune
Phone: (702) 784-4975
Fax: (702) 784-1833
E-Mail: mainofc@seismo.unr.edu
http://www.seismo.unr.edu

Seismological Society of America
201 Plaza Professional Building
El Cerrito, California 94530
Susan Newman
Phone: (510) 525-5474
Fax: (510) 525-7204
E-Mail: info@seismosoc.org
http://www.seismosoc.org/ssa/

SEPM (Society for Sedimentary Geology)
Box 4756
Tulsa, Oklahoma 74159-0756
Executive Director: Cathleen Williams
Phone: (918) 743-9765
Fax: (918) 743-2498
http://julias.ngdc.noaa.gov/mgg/sepm

Eastern Section
Department of Geology
Smith College
Northampton, Massachusetts 01063
H. Allen Curran
Phone: (413) 585-3805
Fax: (413) 585-3786

Great Lakes Section
Illinois State Geological Survey
615 E. Peabody Drive
Champaign, Illinois 61820-6964
C. Pius Weibel
Phone: (217) 333-5108
Fax: (217) 333-2830
E-Mail: weibel@geoserv.isgs.uiuc.edu

Mid-Continent Section
Department of Geology
Kansas State University
Manhattan, Kansas 66506
Allen W. Archer
Phone: (913) 532-6724

North American Micropaleontology Section (NAMS)
23119 Winding Knoll Drive
Katy, Texas 77494-2102
Robert W. Pierce
Phone: (713) 556-2590
Fax: (713) 556-2139

Pacific Section
University of California
Geological Science Dept.
Santa Barbara, California 93106-9630
Cathy Busby
Phone: (805) 893-4068
Fax: (805) 893-2314
E-Mail: cathy@magic.uscsb.edu

Permian Basin Section
Box 1595
Midland, Texas 79702
Paula L. Mitchell
Phone: (915) 683-1573
Fax: (915) 686-7827

Rocky Mountain Section
Box 13947
Denver, Colorado 80201-3947
John Robinson
Phone: (303) 592-4629
Fax: (303) 592-8600
E-Mail: jrobinson@snyderoil.com

Southeastern Section
Department of Geological Sciences
Old Dominion University
Norfolk, Virginia 23829
Diane Kamola
Phone: (804) 683-4301

Sigma Gamma Epsilon
Department of Earth and Physical Sciences
University of Texas at San Antonio
San Antonio, Texas 78249-0663
Eric Swanson
Phone: (210) 691-4455
Fax: (210) 691-4469

Sigma Gamma Epsilon, Epsilon Omega Chapter
Department of Earth and Physical Sciences
University of Texas at San Antonio
San Antonio, Texas 78249-0663
Eric Swanson
Phone: (210) 691-4455
Fax: (210) 691-4469

Society for Archaeological Sciences
Radiocarbon Laboratory
University of California
Riverside, California 92521
R.E. Taylor
Phone: (909) 787-5521
Fax: (909) 787-5409
E-Mail: retaylor@ucrac1.ucr.edu (Bitnet)

Society for Luminescence Microscopy and Spectroscopy
Department of Geological Sciences
University of Tennessee
Knoxville, Tennessee 37996-1410
Kula C. Misra
Phone: (423) 974-2366
Fax: (423) 974-2368
E-Mail: kmisra.rocks@utk.edu

Society for Mining, Metallurgy and Exploration Inc. (SME)
Box 625002
Littleton, Colorado 80162-5002
Gary D. Howell
Phone: (303) 973-9550
Fax: (303) 973-3845
E-Mail: smenet@aol.com
http://www.smenet.

The Society for Organic Petrology
U.S. Geological Survey
12201 Sunrise Valley Drive
956 National Center
Reston, Virginia 20191
c/o Ron Stanton

Society of Economic Geologists
Suite 209
5808 S. Rapp Street
Littleton, Colorado 80120
John A. Thoms
Phone: (303) 797-0332
Fax: (303) 797-0417

Society of Economic Paleontologists and Mineralogists
Gulf Coast Section
165 Pinehurst Road
West Hartland, Connecticut 06091-0065
Bob F. Perkins
Phone: (860) 738-9302
Fax: (860) 738-3542
E-Mail: gcssepm@mail.snet.net
http://www.gcssepm.org

Society of Engineering and Mineral Exploration Geophysicists
Box 10845
Edgemont Branch
Golden, Colorado 80401
R.S. Bell
Phone: (303) 278-9124
Fax: (303) 278-4007

Society of Environmental Geochemistry and Health
Department of Life Sciences
105 B Schrenk Hall
University of Missouri-Rolla
Rolla, Missouri 65401
Paula Lutz
Phone: (314) 341-2205
Fax: (314) 341-4821

Society of Exploration Geophysicists
Box 702740
Tulsa, Oklahoma 74170-2740
John Hyden
Phone: (918) 493-3516
Fax: (918) 493-2074

Society of Independent Professional Earth Scientists
Suite 1106
4925 Greenville Avenue
Dallas, Texas 75206
Diane Finstrom
Phone: (214) 363-1780
Fax: (214) 363-8195

Society of Petroleum Engineers Inc.
Box 833836
Richardson, Texas 75083-3836
Dan K. Adamson
Phone: (214) 952-9393
Fax: (214) 952-9435

Society of Professional Well Log Analysts
Suite C129
6001 Gulf Freeway
Houston, Texas 77023
Vicki J. King
Phone: (713) 928-8925
Fax: (713) 928-9061

Society of Vertebrate Paleontology
401 N. Michigan Avenue
Chicago, Illinois 60611
Pamela D'Argo
Phone: (312) 321-3708
Fax: (312) 245-1085
E-Mail: svp@sba.com
http://eteweb.lscf.uesb.edu/sup/

Soil Science Society of America
677 S. Segoe Road
Madison, Wisconsin 53711
Robert Barnes
Phone: (608) 273-8080
Fax: (608) 273-2021
E-Mail: rbarnes@agronomy.org
http://www.agronomy.org

South Carolina Geological Survey
5 Geology Road
Columbia, South Carolina 29210-4089
C.W. Clendenin
Phone: (803) 896-7708
Fax: (803) 896-7695

South Coast Geological Society Inc.
Box 10244
Santa Ana, California 92711
Diane Murbach
Phone: (619) 275-2474
Fax: (619) 275-1462

South Dakota Geological Survey
Science Center
University of South Dakota
414 East Clark Street
Vermillion, South Dakota 57069-2390
Cleo M. Christensen
Phone: (605) 677-5227
Fax: (605) 677-5895

South Texas Geological Society
Suite D-100
900 N.E. Loop 410
San Antonio, Texas 78209
Doreen Brooner
Phone: (512) 822-9092
Fax: (512) 822-7375

Southeastern Geophysical Society
Halliburton Geophysical Services
Suite 2020
1450 Poydrus Street
New Orleans, Louisiana 70112
Don DuBose
Phone: (504) 592-6781

Southwest Louisiana Geophysical Society
Lafayette, Louisiana 70505
Box 51463, O.C.S.

Student Geological Society
University of Texas
San Antonio, Texas 78249-0663
James O. Jones
Phone: (210) 691-5449
Fax: (210) 691-4469

Tektite Research
University Station, Box X
Austin, Texas 78713-8924
Virgil E. Barnes
Phone: (512) 495-9516
Fax: (512) 471-0140
E-Mail: begmail@begv.beg.utexas.edu
http://www.utexas.edu/research/beg

Tennessee Division of Geology
L&C Tower 13th Floor
401 Church Street
Nashville, Tennessee 37243-0445
Ronald P. Zurawski
Phone: (615) 532-1500

Texas A&M Geological Society
Department of Geology
Texas A&M University
College Station, Texas 77843
Mike Howell
Phone: (409) 845-2451

Texas Bureau of Economic Geology
University of Texas at Austin
University Station, Box X
Austin, TX 78713-8924
Noel Tyler
Phone: (512) 471-1534
Fax: (512) 471-0140
E-Mail: begmail@begv.beg.utexas.edu
http://www.utexas.edu/depts/beg/

Texas Water Resources Institute
Texas A&M University

College Station, Texas 77843-2118
W.R. Jordan
Phone: (409) 845-1851
Fax: (409) 845-8554
E-Mail: twri.tamu.edu
http://twri.tamu.edu

Tobacco Root Geological Society

Box 2734
Missoula, Montana 59806
Marie Marshall
Phone: (406) 496-4327

Tulsa Geological Society Inc.

Box 4508
Tulsa, Oklahoma 74159-0508
Jean R. Lemmon
Phone: (918) 582-4762

U.S. Antarctic Marine Geology Research Facility

108 Carraway Bldg.
Florida State University
Tallahassee, Florida 32306-3026
Thomas Janecek
Phone: (904) 644-2407
Fax: (904) 644-4214
E-Mail: curator@gly.fsu.edu
http://www.gly.Fsu.edu/curator/index.html

U.S. Antarctic Research Program

Polar Information Program
National Science Foundation
4201 Wilson Boulevard
Arlington, Virginia 22230
Cornelius Sullivan
Phone: (703) 306-1030

U.S. Bureau of the Census

Public Information Office
Room 2705-3
Washington, DC 20233
Phone: (301) 457-2800
Fax: (301) 457-3670
E-Mail: pio@census.gov
http://www.census.gov

U.S. Committee on Irrigation and Drainage

Suite 483
1616 17th Street
Denver, Colorado 80202
Larry D. Stephens
Phone: (303) 628-5430
Fax: (303) 628-5431
E-Mail: stephens@uscid.org
http://www.uscid.org/~uscid

U.S. Department of Agriculture -- Forest Service

Minerals and Geology Management
14th Street and Independence Avenue, S.W.
Washington, D.C. 20090-6090
Phone: (202) 205-1224
Fax: (202) 205-1243

U.S. Department of Commerce

International Trade Administration, Basic Industries
Office of Materials, Machinery, and Chemicals
14th and Constitution Ave. N.W.
Washington, D.C. 20230
Phone: (202) 482-0575

U.S. Department of Defense

Defense Logistics Agency
Planning and Market Research
1745 Jefferson Davis Highway
Arlington, Virginia 22202
Phone: (703) 607-3202
(202) 586-8800

U.S. Department of Energy

Forrestal Building
1000 Independence Avenue, S.W.
Washington, D.C. 20585
Phone: (202) 586-4670
Fax: (202) 586-5049

National Energy Information Center
Energy Information Administration
Forrestal Building, Room 1F-048
Washington, D.C. 20585
Phone: (202) 586-8800
E-Mail: infoctr@eia.doe.gov
http://www.eia.doe.gov

U.S. Department of the Interior

1849 C Street, N.W.
Washington, D.C. 20240
Secretary: Bruce Babbitt
Phone: (202)208-3171
Fax: (202)208-6950

Bureau of Indian Affairs
Office of Trust Responsibilities
Division of Energy and Mineral Resources
Room 239
730 Simms Street
Golden, Colorado 80401
Richard N. Wilson
Phone: (303) 231-5070
Fax: (303) 231-5085
http:snake2.cr.usgs.gov

Bureau of Land Management
Office of Public Information

MS 5600
1849 C Street, N.W.
Washington, D.C. 20240
Phone: (202) 208-5717

Bureau of Reclamation
Office of Public Information
MS 7642
1849 C Street, N.W.
Washington, D.C. 20240
Phone: (202) 208-4662

Office of Surface Mining Reclamation & Enforcement
Office of Communications
Room 262
1951 Constitution Avenue, N.W.
Washington, D.C. 20241
Phone: (202) 208-2565

U.S. Department of Labor

Bureau of Labor Statistics
Office of Publications
Postal Square Building
2 Massachusetts Avenue, N.E.
Washington, D.C. 20212-0001
Phone: (202) 606-5886

Mine Safety and Health Administration
Office of Information and Public Affairs
Room 601
4015 Wilson Boulevard
Arlington, Virginia 22203
Phone: (703) 235-1452

U.S. Environmental Protection Agency

401 M Street, S.W.
Washington, D.C. 20460
Administrator: Carol M. Browner
Phone: (202) 260-2080

Prevention, Pesticides, and Toxic Substance
401 M. Street, S.W.
Washington, D.C. 20460
Phone: (202) 260-2902

Solid Waste and Emergency Response
401 M. Street, S.W.
Washington, D.C. 20460
Phone: (202) 260-4610

Water
401 M Street, S.W.
Washington, D.C. 20460
Phone: (202) 260-7400

U.S. Geodynamics Committee

National Academy of Sciences
2101 Constitution Avenue, N.W.
Washington, D.C. 20418
Charles Meade
Phone: (202) 334-2744
Fax: (202) 334-1377.
E-Mail: cmeade@nas.edu

U.S. Geological Survey (USGS) (C,E,G,H)

12201 Sunrise Valley Drive
National Center
Reston, Virginia 20191
Director: Dr. Gordon P. Eaton
Phone: (703) 648-7411
Fax: (703) 648-4454
http://www.usgs.gov/

Saudi Arabian Mission
Unit 62101
APO AE 09811-2101
Phone: (011-966)2 667-4188
Fax: (011-966)2 660-5624

U.S. International Trade Commission

Division of Minerals and Metals
500 E. Street, S.W.
Washington, D.C. 20436
Phone: (202) 205-3419

U.S. National Committee for Geology

Mail Stop 917
U.S. Geological Survey
Reston, Virginia 20191
Paul P. Hearn Jr.
Phone: (703) 648-6287
Fax: (703) 648-4227
E-Mail: phearn@usgs.gov

Utah Geological Association

Box 11334
Salt Lake City, Utah 84147
Lori C. Robison
Phone: (801) 521-9255
Fax: (801) 521-0380

Utah Geological Survey

Box 146100
1594 West North Temple, Suite 3410
Salt Lake City, Utah 84114-6100
M. Lee Allison
Phone: (801) 537-3300
Fax: (801) 537-3400.
E-Mail: lallison@email.state.ut.us
http://utstdpwww.state.ut.us/~ugs

Vermont Agency of Natural Resources

Vermont Geological Survey
Center Building
103 S. Main Street
Waterbury, Vermont 05671-0301
Laurence R. Becker
Phone: (802) 241-3496
Fax: (802) 244-1102
E-Mail: larryb@anrimsgis.anr.state.vt.us

Vermont Geological Society
Department of Geology
University of Vermont
Burlington, Vermont 05405-0122
Stephen Wright
Phone: (802) 656-4479
Fax: (802) 656-0045
E-Mail: swright@moose.uvm.edu

Virginia Division of Mineral Resources
Box 3667
Charlottesville, Virginia 22903
Stanley S. Johnson
Phone: (804) 293-5121
Fax: (804) 293-2239

Washington Department of Natural Resources
Division of Geology and Earth Resources
Box 47007
Olympia, Washington 98504-7007
Raymond Lasmanis
Phone: (360) 902-1450
Fax: (360) 902-1785
E-Mail: cjmanson@u.washington.edu

Water Environment Federation
601 Wythe Street
Alexandria, Virginia 22314-1994
Phone: (703) 684-2400
Fax: (703) 684-2492

West Texas Geological Society Inc.
Box 1595
Midland, Texas 79702
Paula L. Mitchell
Phone: (915) 683-1573
Fax: (915) 686-7827

West Virginia Coal Mining Institute Inc.
Box 6070
West Virginia University
Morgantown, West Virginia 26506-6070
Royce J. Watts
Phone: (304) 293-5695
Fax: (304) 293-5708
E-Mail: Watts@wvucomer1.comer.wvu.edu

West Virginia Geological and Economic Survey
Mont Chateau Research Center
Box 879
Morgantown, West Virginia 26507-0879
Larry D. Woodfork
Phone: (304) 594-2331
Fax: (304) 594-2575
E-Mail: info@geoserv.wvnet.edu

Wisconsin Geological and Natural History Survey
3817 Mineral Point Road
Madison, Wisconsin 53705-5100
James M. Robertson
Phone: (608) 262-1705
Fax: (608) 262-8086

Women in Mining
Suite 400
1801 Broadway
Denver, Colorado 80202
Karen Grass
Phone: (303) 298-1535

World Data Center A: Glaciology (Snow and Ice)
CIRES
University of Colorado
Campus Box 449
Boulder, Colorado 80309
Roger G. Barry
Phone: (303) 492-6199
Fax: (303) 492-2468
E-Mail: nsidc@kryos.colorado.edu

World Data Center A: Marine Geology and Geophysics
National Oceanic and Atmospheric Administration
Mail Code E/GC3
325 Broadway
Boulder, Colorado 80303
Michael S. Loughridge
Phone: (303) 497-6487
Fax: (303) 497-6513

World Data Center A: Oceanography
National Oceanic and Atmospheric Administration
Room 409
1825 Connecticut Avenue, N.W.
Washington, D.C. 20235
Ronald E. Moffatt
Phone: (202) 673-5571

World Data Center A: Solid Earth Geophysics
NGDC/NESDIS/NOAA
325 Broadway
Boulder, Colorado 80303
Allen M. Hittelman
Phone: (303) 497-6521
Fax: (303) 497-6513
E-Mail: amh@ngdc.noaa.gov
http://www.ngdc.noaa.gov

Wyoming Geological Association
Box 545
Casper, Wyoming 82602

Phone: (307) 237-0027

Wyoming State Geological Survey
Box 3008
University Station
Laramie, Wyoming 82071
Gary B. Glass
Phone: (307) 766-2286
Fax: (307) 766-2605
E-Mail: wsgs@wsgs.uwyo.edu
http://www_wwrc.uwyo.edu/wrds/wsgs/wsgs.html

URUGUAY

Dirección Nacional de Topografia (C)
Ministerio de Transporte y Obras Publicas
Rincón 575, Piso 3
Montevideo 11000
Director: Ricardo Bértolo
Phone: (598-2)959434
Fax: (598-2)952673
E-Mail: uy33786@antel.com.uy

Direccion Nacional de Mineria y Geologia (DINAMIGE) (G)
Hervidero 2861
Montevideo
National Director: Dr. Diego Diana Montorell
Phone: (598-2)293196/293473
Fax: (598-2)293196
Telex: 22072 MINIE UY

Geological Institute of Uruguay
Instituto Geologico del Uruguay
Calle Hervidero 2853
Montevideo
Uruguay

Instituto Nacional de Pesca (INAPE) (G)
Constituyente 1497
Montevideo
General Director: Captain de Navio
Phone: (598-2)492969

Servicio de Oceanografia, Hidrografia y Meteorologia de la Armada (SOHMA) (C)
Capurpo 980
Montevideo
Director: Capitan de Navio
Phone: (598-2)399220

Servicio Geografico Militar (C)
Avenida 8 de Octubre 3255
Montevideo
Director: Coronel Ivho R. Acuna

Phone: (598-2)807111

UZBEKISTAN

State Committee on Geology (G,R)
11 Shevchenko St.
Tashkent 700060
Chairman: Tulkun Shayakubor
Phone: (3-712)337206
Fax: (3-712)560283
Telex: 116108

VANUATU

Department of Geology (G)
Mines and Water Resources
Private Mail Bag 001
Port Villa
Phone: 22423
Fax: 22213

VENEZUELA

Consejo Consultivo de Directores de Servicios Geologicos de Latinoamerica (G)
Torre Oeste, Piso 5
Parque Central
Caracas

CVG - Tecnica MInera, C.A. (TECMIN) (G)
C.C. Chilemex
Avenida Estados Unidos cruce con Calle México
Piso 1, Locales 4, 5, y 6
Urb. Chilemex
Puerto Ordaz-Edo. Bolivar
President: Jesús Rebolledo
Phone: (58-86)228377/223439
Fax: (58-86)220729
Telex: 86249 FADM VC

Direccion General Sectorial de Minas y Geologia (G)
  Department of Geology
  Director: Dr. Emilio Herrero Olivares
  Phone: (58-2)5075414/5075415
  Telex: 22594/21692 MEM VC

  Direccion de Minas
  Director: Dra. Sara Barrios de Rodriuez

Venezuela

Phone: (58-2)5075601/5602
Telex: 22594/21692 MEM VC

Ministerio de Energia y Minas
Torre Oeste, Piso 4
Parque Central
Caracas 1010
Oswaldo Ruiz
Phone: (58-2)5075401
Fax: (58-2)5752497

Fundacion Venezolana de Investigaciones Seismologicas (FUNVISIS) (G)
Prolongacion Calle Mara
Urbanizacion El Llanito
Caracas
Presidente: Dr. Cesar Hernandez Acosta
Phone: (58-2)385416/5894/5417/5053

Service Autonoma de Geoghrafica y Cartografia Nacional (C)
Edificio Camejo, Piso 2, Ofic. 231
Centro Simon Bolivar
Caracas 1010
Director: Aliocia Moreau D.
Phone: (58-2)4081710/1711
Fax: (58-2)5450374
E-Mail: sagecan@conicit.ve

Venezuelan Geological Society
Sociedad Venezolana de Geologos
Apartado 2006
Caracas 1010A
Anibal R. Martinez
Phone: (58-2) 572 0734
Fax: (58-2) 606 4963

Venezuelan Society for the History of Geological Sciences
Sociedad Venezolana de Historia de las Geociencias
Apartado 47334
Caracas 1041A
Andre Singer
Phone: (58-2) 2575153
Fax: (58-2) 2575153
E-Mail: dptoct@funvisis.internet.ve

Venezuelan Society of Petroleum Engineers
Sociedad Venezolana de Ingenieros de Petroleo
c/o Colegio de Ingenieros de Venezuela
Apartado 2006
Caracas

Venezuelan Speleological Society
Sociedad Venezoland de Espeleologia
Apartado 47334
Caracas 1041A

Carlos Bosque
Phone: (58-2) 746436
Fax: (58-2) 746436
E-Mail: carlosb@usb.ve

VIET NAM

Geological Survey of Viet Nam (G)
6 Pham Ngu Lao Street
Hanoi
Deputy General Director: Dr. Tran Van Tri
Phone: (84-4)82 60674
Fax: (84-4)82 54734

Institute of Geophysics (C)
Vietnam National Centre for Natural Science and Technology
Box 411
Buu Dien Bo Ho
Hanoi
Deputy Director: Nguyen Thi Kim-Thoa
Phone: (84-4)8352380
Fax: (84-4)8364696
Telex: 411525 NCSR VT
E-Mail: thoa@igp.ac.vn

Institute of Geosciences
Vien Dia Chat
NCNS&T of Viet Nam
Nghia Do - Tu Liem - Hanoi
Box 603
10.000 Viet Nam
Director: Nguyen Trong Yem
Phone: (84-4) 8351492
Fax: (84-4) 8362886

Institute of Oil and Gas Research
Hai Hung Province
Hung Yen

Mining Association
54 Hai Ba Trung
Hanoi

Research Institute of Geology and Mineral Resources (G)
Thanh Xuan
Dong Da
Hanoi
Director: Phan Cu Tien
Phone: (84-4)8542123
Fax: (84-4)8542125

Service Geographique Nationale (C)
Ministere de la Defense Nationale
Dalat
Director: Nguyen Van Khai

Viet Nam

Service Hydrometeorologique (H)
4 rue Dang Thai Than
Hanoi
Director General: Tran Van An

YEMEN

Central Planning Organization (G)
P.O. Box 175
Sanaa
Chairman: Motahar A. Al-Saeede
Phone: (967-2)250117

Department of Hydrology (YOMINCO) (H)
Sanaa
Director General: Ali Jaber Alawi
Phone: (967-2)207040
Telex: 2257 MAHROKAT YE

Mineral Exploration Board (R)
Mineral Resources Sector
Sanaa
Chairman: Ali Jabr Alawi
Phone: (967-2)252255
Fax: (967-2)251624
Telex: 3310MEB

Ministry of Oil and Mineral Resources (G)

Sanaa
Director General: Abdul Rehman K. Mohammed
Phone: (967-2)206819Ye

Geological Survey Department
Director General: Abdutawab Kaid

Yemen Geological Society (E,G)
P.O. Box 11795
Sanaa
President: Khalid Ahmed Al-Suba'i
Phone: (967-1)224981

Yemen Oil and Mineral Resources Corporation (G)
P.O. Box 81
Sanaa
Minister: E.H. Ahmed Ali Al-Mohani
Phone: (967-2)202313
Telex: 2257 MAHROKAT YE

YUGOSLAVIA

Federal Administration for International Scientific, Educational, Cultural and Technical Cooperation (FAISECTC) (H)
29, Kosancicev venac
11000 Belgrade
Director General: Miljenko Zrelec
Phone: (38-11)625955

Federal Hydrometeorological Institute (H)
6, Bircaninova
P.O.B. 604
11000 Belgrade
Director: Mile Sikic
Phone: (38-11)646555

Geoinstitut
Rovinjska 12, 11000 Belgrade
Dimitrije Cvetkovic
Phone: (38-11) 4889966
Fax: (38-11) 4885296

Institute for Cartography (C)
39, Bulevar Vojvode Misica
11000 Belgrade
Phone: (38-11)651255

Institute for Development of Water Resources (H)
80, Jaroslava Cernija
11000 Belgrade
Director General: Milorad Miloradov
Phone: (38-11)649265

Institute for Geological, Geophysical and Mining Exploration of Nuclear and other Mineral Resources (G)
12, Rovinjska
11000 Belgrade
Director General: Radule Popovic
Phone: (38-11)489966
General Director for Research Coordination: Vladimir Saric
Phone: (38-11)489966

Institute for Geological and Geophysical Research (G)
48, Karadjordjeva
11000 Belgrade
Director General: Petar Vojnovic
Phone: (38-11)625931

Institute for Hydrogeology and Hydrotechnology (G,H)
3, Ustanicka

71000 Sarajevo
Director General: Stanko Celikovic
Phone: (38-71)625833

Organization for Complex Geological Research (G)
88a, Savska Cesta
P.O.B. 334
41001 Zagreb 2
Director: Jelisaveta Molnar

ZAIRE

Department of Geophysics
C.R.S.N.
Lwiro
D.S. Bukavu (Sud-Kivu)
Zana Ndontoni

Department of Seismology
I.R.S.
Lwiro
D.S. Bukavu (Sud-Kivu)

Geological and Mining Research Center (G)
Ministry of Scientific Research and Technology
44 Avenue des Huileries
B.P. 898
Kinshasa
Director: Monama Ondongo
Phone: (243-12)22944

Zairian Geographic Institute (C)
106 Blvd du 30 Juin
B.P. 3086
Kinshasa
Director: Lt. Col. Suani-di-Naba
Phone: (243-12)31854/31039

Zairian Riverways Authority (H)
B.P. 11697
Kinshasa
Director General: Mualu Kitenda
Phone: (243-12)22704/26526/24471

ZAMBIA

Geological Survey Department (G)
Ministry of Mines and Minerals Development
P.O. Box 50135
Lusaka
Director: D. Mulela

Phone: (260-1)251655
Fax: (260-1)251973
Telex: 40107 ZA

Mining Industry and Technical Services (G)
Zambia Consolidated Copper Minesa Limited (ZCCM)
P.O. Box 10
Kalulushi
Manager: Dr. Silane Mwenechanya

Survey Department (C)
Ministry of Lands and Natural Resources
Mulungushi House
P.O. Box 50397
Lusaka
Surveyor General: Andrew J. Kownacki

Water Affairs Department (H)
Ministry of Agriculture and Water Development
Mulungushi House
P.O. Box 50288
Lusaka
Director: C.R.W. Kayombo

University of Zambia
School of Mines
Geology Department
Box 32379
Lusaka
Phone: (260-1) 294318
Fax: (260-1) 253952
http://www.inza.zm/newpage/min.html

Zambia Industrial and Mining Corporation, Limited (ZIMCO) (G)
Zimco House, Cairo Road
P.O. Box 30090
Lusaka
Director General: James Mapoma

ZIMBABWE

Department of Geological Survey (G)
Ministry of Mines
Maufe Bldg., Selous Ave./5th St.
P.O. Box CY210, Causeway
Harare
Acting Director: Wishes Magalela
Phone: (263-4)726344
Fax: (263-4)739601
Telex: 22416 MINES ZW

Department of the Surveyor General (C)

Department of the Ministry of Lands and Water Development
P. O. Box CY540, Causeway
Harare
Phone: (263-4)794545
Fax: (263-4)794540

Hydrological Branch (H)

Division of Water Resources and Development
P.O. Box 8132, Causeway
Harare
Chief: W.G. Wannell
Phone: (263-4)707861

Institute of Mining Research

University of Zimbabwe
Box MP 167
Mount Pleasant, Harare
Phone: (263-4) 303211
Fax: (263-4) 303292

Zimbabwe School of Mines

Box 1392
Bulawayo
Phone: (263-9) 71246/71247

ORGANIZATION INDEX

Academia Sinica (Chinese Academy of Sciences), China	23
Academy of Scientific Research and Technology, Egypt	36
Administration du Cadastre et de la Topographie, Luxembourg	64
Aerial Survey of Egypt, Egypt	36
Afghan National Petroleum Company, Afghanistan	1
Agency for the Assessment and Application of Technology (BPPT), Indonesia	49
Agency of Natural Resources and Energy, Japan	55
Agricultural Office, Liechtenstein	63
Agricultural Research Center, Libyan Arab Jamahiriya	63
Agriculture and Forestry Aerial Survey Institute, Taiwan, Republic Of China	96
Agriculture and Resource Management Council of Australia and New Zealand, Australia	3
Aichi University of Education, Japan	55
Akita University, Japan	55
Alabama Geological Society, United States	108
Alaska Division of Geological and Geophysical Surveys, United States	108
Alaska Geological Society, United States	108
Albanian Geological Survey, Albania	1
Alberta Department of Energy, Canada	17
Alberta Geological Survey, Canada	17
Albuquerque Geological Society, United States	109
Alfred-Wegener-Stiftung, Germany	41
All-Russia Petroleum Scientific Research Geological Exploration (VNIGRI), Russian Federation	82
All-Russian Geological Petroleum Exploration Research Institute (VNIGNI), Russian Federation	82
All-Russian Research Institute for Geology and Mineral Resources, Russian Federation	82
All-Russian Research Institute for Hydrology & Engineering Geology (VSEGINGEO), Russian Federation	82
All-Russian Scientific Research Institute of Economics of Mineral Resources and Mineral Lands Use (VIEMS), Russian Feder	82
All-Union Geological Research Institute, Russian Federation	82
Amarillo Geophysical Society, United States	109
American Association for Zoological Nomenclature, United States	109
American Association of Petroleum Geologists, United States	109
American Association of Stratigraphic Palynologists, United States	109
American Chemical Society/Geochemistry Division, United States	109
American Congress on Surveying and Mapping, United States	109
American Crystallographic Association, United States	109
American Geological Institute, United States	109
American Geophysical Union, United States	109
American Ground Water Trust, United States	110
American Institute of Hydrology, United States	110
American Institute of Professional Geologists, United States	110
American Lunar Society, United States	110
American Quaternary Association, United States	110
American Society for Photogrammetry and Remote Sensing, United States	110
American Water Resources Association, United States	110
Ammosov Yakutsk State University, Russian Federation	83

ORGANIZATION INDEX

AMUR Integrated Research Institute, Russian Federation	83
Apia Observatory, Samoa	87
Appalachian Geological Society, United States	110
Applied Geology Department, Myanmar	70
Arab Center for the Study of Arid Zones and Dry Lands (ACSAD), Syrian Arab Republic	96
Arab Geologist Association, Iraq	51
Arctic Research Commission, United States	110
Argentine Association of Mineralogy, Petrology and Sedimentology, Argentina	2
Argentine Geological Association, Argentina	2
Argentine Geological Survey, Argentina	2
Argentine Paleontological Association, Argentina	2
Arizona Geological Society, United States	110
Arizona Geological Survey, United States	110
Arkansas Geological Commission, United States	110
Arrondissement Minéralogique de la Guyane, Guadeloupe	43
Arthur Holmes Society, United Kingdom	103
Arthur Lakes Library, United States	110
ASFE/Professional Firms Practicing in the Geosciences, United States	110
Association for Women Geoscientists, United States	111
Association Nationale des Services d'Eau (ANSEAU), Belgium	11
Association of African Geological Surveys, Morocco	69
Association of American State Geologists, United States	111
Association of Arab Geologists, Iraq	51
Association of Australasian Paleontologists, Australia	4
Association of Earth Science Editors, United States	111
Association of Engineering Geologists, United States	111
Association of Exploration Geochemists, Canada	17
Association of Exploration Geochemists, United States	111
Association of Geomorphologists of Turkey, Turkey	101
Association of Geoscientists for International Development, Brazil	13
Association of Ground Water Scientists and Engineers, United States	111
Association of Italian Geophysicists, Italy	53
Association of Professional Engineers, Geologists and Geophysicists of Alberta (APEGGA), Canada	17
Association of Professional Geologists and Geophysicists of Quebec, Canada	17
Association of Scientific Workers of the People's Republic of Albania, Albania	1
Association of Spanish Geologists, Spain	91
Association of Spanish Petroleum Geologists and Geophysicists, Spain	92
Association of Women Soil Scientists, United States	111
Athlone Regional Technical College, Ireland	51
Atlanta Geological Society, United States	111
Atlantic Coastal Plain Geological Association, United States	111
Atlantic Geoscience Centre, Canada	17
Atlantic Geoscience Society, Canada	17
Austin Geological Society, United States	111
Australasian Institute of Mining and Metallurgy, Australia	4
Australian Geological Survey Organisation, Australia	4
Australian Geomechanics Society, Australia	4
Australian Geoscience Information Association Inc., Australia	4
Australian Institute of Geoscientists, Australia	4
Australian Mineral Foundation Inc., Australia	4
Australian National University, Australia	4

ORGANIZATION INDEX

Australian Society of Exploration Geophysicists, Australia	4
Austrian Geological Society, Austria	9
Austrian Mineralogical Society, Austria	9
Austrian Paleontological Society, Austria	9
Ballarat University College, Australia	4
Bangladesh Geological Society, Bangladesh	10
Bangladesh Meteorolgical Department, Bangladesh	10
Bangladesh Oil and Gas Development Corp. (PETROBANGLA), Bangladesh	10
Bangladesh Water Development Board, Bangladesh	10
Bashkortostan State University, Russian Federation	83
Baton Rouge Geological Society, United States	111
Bay Area Geophysical Society, United States	111
Baylor Geological Society, United States	111
Beijing Natural History Museum, China	25
Beijing Research Institute of Uranium Geology, China	25
Beijing University of Science and Technology, China	25
Belgian Contact Group on Clays, Belgium	11
Belize Meteorologist/Hydrologist, Belize	11
Ben-Gurion University of the Negev, Israel	52
Federal Institute for Geosciences and Natural Resources (Bundesanstalt fur Geowissenschaften und Rohstoffe), Germany	42
Big Rivers Area Geological Society, United States	111
Birkbeck College, United Kingdom	103
Board on Infrastructure and the Constructed Environment, United States	111
Board on Natural Disasters and U.S. National Committee for the Decade for Natural Disaster Reduction, United States	112
Brazilian Geological Society, Brazil	13
Brazilian Geophysical Society, Brazil	13
British Antarctic Survey, United Kingdom	103
British Columbia Geological Survey Branch, Canada	17
British Geological Survey, United Kingdom	103
British Geotechnical Society, United Kingdom	103
British Micropalaeontological Society, United Kingdom	104
Bulgarian Academy of Sciences, Bulgaria	15
Bureau de Recherches Geologiques et Minieres (BRGM), Gabon	41
Bureau de Recherches Geologiques et Minieres, Niger	74
Bureau for Cartography and Geology, Senegal	87
Bureau for Geological and Mineral Research (BRGM), Senegal	87
Bureau of Geological and Mineral Research, French Guiana	41
Bureau of Geological Concessions, Poland	79
Bureau of Geology and Mining, China	25
Bureau of Mines and Geology of Burkina, Burkina Faso	16
Bureau of Mines and Geology, Senegal	88
Bureau of Mines, Liberia	63
Bureau of Petroleum and Marine Geology, China	25
Bureau of Water Resources, Korea, Republic Of	61
Burkina Geographic Institute, Burkina Faso	16
Cadastral Direction, Burkina Faso	16
Cadastre du Congo, Congo	31
California Division of Mines and Geology, United States	112
California Earthquake Society, United States	112
California Groundwater Association, United States	112

137

ORGANIZATION INDEX

Canadian Continental Drilling Program, Canada	17
Canadian Geological Survey	17
Canadian Geological Survey — Minerals and Regional Geoscience Branch	18
Canadian Geological Survey — Sedimentary and Marine Geoscience Branch	18
Canadian Geophysical Union, Canada	18
Canadian Geoscience Council, Canada	18
Canadian Geotechnical Society, Canada	18
Canadian Institute of Mining, Metallurgy, and Petroleum, Canada	18
Canadian Quaternary Association, Canada	19
Canadian Society of Petroleum Geologists, Canada	19
Canadian Society of Soil Science, Canada	19
Canadian Well Logging Society, Canada	19
Carolina Geological Society, United States	112
Carpathian Balkan Geological Association, Poland	80
Carpathian-Balkan Geological Association Commission on Stratigraphy, Paleogeography and Paleontology, Hungary	46
Cartographic Department, Finland	38
Cartography Service, Netherlands	71
Center for Geotechnical Research, El Salvador	37
Center for the Earth Sciences, United States	112
Center for Volcanological Research, France	39
Center of Geology, Institute of Tropical Scientific Research, Portugal	80
Center of Integrated Surveys of Natural Resources by Remote Sensing (CLIRSEN), Ecuador	35
Centraal Bureau Luchtkaartering, Suriname	94
Central African Mineral Resources Development Centre, Congo	31
Central Fuel Research Institute, India	48
Central Geological Bureau of Hungary, Hungary	46
Central Geological Survey, Taiwan, Republic Of China	96
Central Ground Water Board, India	48
Central Hydrographic Service, Italy	53
Central Institute for Meteorology and Geodynamics, Austria	9
Central Laboratory for Geodesy, Bulgaria	15
Central Laboratory for Space Research, Bulgaria	15
Central Laboratory of Mineralogy and Crystallography, Bulgaria	15
Central Mining Institute, Poland	80
Central Mining Research Station, India	48
Central Office of Geology, Poland	80
Central Office of the National Land Survey, Sweden	94
Central Planning Organization, Yemen	132
Central Research Institute of Geological Prospecting for Base & Precious Metals (TSNIGRI), Russian Federation	83
Central Research Organization, Myanmar	70
Central South University of Technology, China	26
Central Survey and Mapping Agency, Canada	19
Central Water Authority, Mauritius	67
Central Water Commission, India	48
Centre de Cartographie et de Télédétection de la DCGTx (DCGTx/CCT), Cote D'ivoire	32
Centre Européen de Géodynamique et de Seismologie, Luxembourg	64
Centre for Ore Deposit and Exploration Studies (CODES), Australia	4
Centre National de la Recherche Appliquee, Benin	12
Centre National de la Recherche Scientifique (CNRS), France	39
Centre National de la Recherche Scientifique et Technique, Benin	12

ORGANIZATION INDEX

Centre National de Recherche et d'Applications des Geosciences, Algeria	1
Centre National de Recherches Oceanographiques, Madagascar	64
Centre National des Zones Arides, Algeria	1
Centre of Geoscience Research, Slovakia	89
Centre Royal de Teledetection Spatiale, Morocco	69
Centro de Estudios Hidrograficos, Spain	92
Centro de Investigacion Cientifica, Mexico	67
Centro de Investigaciones Geologicas, Bolivia	12
Centro de Investigaciones Minero-Metalurgicas, Chile	23
Centro de Tecnologia Petrolera, Bolivia	12
Chamber of Geological Engineers of Turkey, Turkey	101
Chamber of Geophysical Engineers of Turkey, Turkey	101
Charles Darwin Research Station, Ecuador	35
Chengdu Institute of Technology, China	26
Chernyshevsky Saratov State University, Russian Federation	83
Chiang Mai University, Thailand	98
Chiba University, Japan	55
Chilean Association of Seismology and Earthquake Engineering, Chile	23
China Association of Geological Education, China	26
China Institute of Hydrogeology and Engineering Geology Exploration, China	26
China National Coal Corporation, China	26
China National Geological Exploration Centre of Building Materials Industry, China	26
China National Nonferrous Metals Industry Corporation, China	26
China National Nuclear Corporation, China	26
China National Offshore Corporation, China	26
China National Petroleum Corp., China	26
China Ocean Mineral Resources Association, China	27
China University of Geosciences, China	26
China University of Geosciences, Wuhan, China	26
China University of Mining and Technology, China	26
Chinese Academy of Geoexploration, China	27
Chinese Academy of Geological Sciences (CAGS), China	27
Chinese Academy of Surveying and Mapping, China	27
Chinese Committee of Geothermal Exploitation Management, China	27
Chinese Institute of Geology and Mineral Resources Information, China	27
Chinese National Nuclear Industry Company, China	27
Chita Polytechnical Institute, Russian Federation	83
Chulalongkorn University, Thailand	98
Ciudad Universitaria, Paraguay	77
Club de Mineralogie de Montreal Inc., Canada	19
Coal Research Limited, New Zealand	72
Coast and Geodetic Survey Department, Philippines	79
Coastal Bend Geophysical Society, United States	112
Colegio de Ingenieros Geólogos de México, A.C., Mexico	67
College of Integrated Arts and Sciences, Japan	56
Colombian Volcano Observatory, Colombia	31
Colorado Geological Survey, United States	112
Colorado Water Congress, United States	112
Comisíon Federal de Electricidad, Mexico	67
Comisíon Nacional del Agua, Mexico	67
Commission for the Geological Map of the World, France	39
Commission on Geographic Information Systems, Costa Rica	32

ORGANIZATION INDEX

Commission on Glaciation (INQUA), United States	112
Commission on Loess, Russian Federation	83
Commission on Quaternary of Africa, Senegal	88
Commission on the Quaternary of South America, Argentina	2
Committee of Geology and Mineral Resources, Bulgaria	15
Committee on Coral Reefs, United States	112
Committee on Solid Earth Sciences, Russian Federation	83
Commodity Futures Trading Commission, United States	112
Commonwealth Scientific and Industrial Research Organization (CSIRO), Australia	5
Companhia de Pesquisa de Recursos Minerais (CPRM), Brazil	13
Compania Nacional Productora de Cemento (CANAC), Nicaragua	73
Connecticut Geological and Natural History Survey, United States	112
Consejo Consultivo de Directores de Servicios Geologicos de Latinoamerica, Venezuela	130
Consejo de Recursos Minerales, Mexico	67
Consejo Nacional de Ciencia y Tecnologia, Mexico	67
Consejo Nacional de Investigaciones Cientificas y Technologicas (CONICIT), Costa Rica	32
Conservation and Survey Division, United States	113
Continental Geoscience Division, Canada	19
Coordinacion de Hidrologica, Spain	92
Coordinating Committee for Coastal and Offshore Geoscience Programmes in East and Southeast Asia (CCOP), Thailand	98
Coordinating Committee on the Himalayan Region, Finland	38
Coordination and Planning Division, Canada	19
Coordinative and Administrative Body, Poland	80
Corporacion Nicaraguense de Minas (INMINE), Nicaragua	73
Council on Undergraduate Research (Geology Division), United States	113
Coventry University, United Kingdom	104
Curtin University of Technology, Australia	5
Cushman Foundation for Foraminiferal Research, United States	113
CVG - Tecnica MInera, C.A. (TECMIN), Venezuela	130
Czech Geological Survey, Czech Republic	34
Dallas Geological Society, United States	113
Dallas Paleontological Society, United States	113
Dallas Society of Potential Field Geophysicists, United States	113
Danish Land Development, Denmark	34
Defense Mapping Agency (DMA), United States	113
Delaware Geological Survey, United States	113
Denver Geophysical Society, United States	113
Departamento de Ciencias Geograficas, Dominican Republic	35
Departamento de Desarrollo Geologico y Recursos Minerales, Costa Rica	32
Departamento de Geologia, Spain	92
Departamento Nacional de Obras de Saneamento (DNOS), Brazil	13
Departamento Nacional de Producao Mineral (DNPM), Brazil	13
Departement Hydro-Geologie Ministere des Mines de l'Energie, des Resources Hydrauliques et des Travaux Publics, Togo	100
Department des Recherches Agronomiques, Benin	12
Department of Earth Sciences, Taiwan, Republic Of China	96
Department of Energy, Mines, and Petroleum Resources, Canada	19
Department of Energy, Philippines	79
Department of Environmental Engineering, Taiwan, Republic Of China	96
Department of Environmental Protection, Taiwan, Republic Of China	97
Department of Geodesy and Realty Registration (DLV), Aruba	3

ORGANIZATION INDEX

Department of Geography and Geology, Hong Kong	45
Department of Geological Survey and Exploration, Myanmar	70
Department of Geological Survey, Zimbabwe	133
Department of Geology and Mines, Bhutan	12
Department of Geology and Mines, Ecuador	35
Department of Geology and Mines, Lao People's Democratic Republic	62
Department of Geology, Kenya	60
Department of Geology, Vanuatu	130
Department of Geophysics and Volcanology, Italy	53
Department of Geophysics, Zaire	133
Department of Hydrology (YOMINCO), Yemen	132
Department of Indian and Northern Affairs Canada, Canada	19
Department of Industrial Development, Qatar	81
Department of Irrigation and Drainage, Malaysia, Malaysia	65
Department of Irrigation and Hydrology, Nepal	71
Department of Lands and Surveys, Cyprus	33
Department of Lands and Surveys, Gambia	41
Department of Lands and Surveys, Jordan	60
Department of Lands and Surveys, Uganda	102
Department of Lands, Surveys and Physical Planning, Lesotho	62
Department of Marine Resources, Taiwan, Republic Of China	97
Department of Mineral Resources, Australia	5
Department of Mineral Resources, Tanzania, United Republic Of	98
Department of Mineral Resources, Thailand	98
Department of Minerals and Energy, Australia	5
Department of Mines and Energy South Australia, Australia	5
Department of Mines and Energy, Canada	19
Department of Mines and Geology, Afghanistan	1
Department of Mines and Geology, Cameroon	16
Department of Mines and Geology, Nepal	71
Department of Mines, Botswana	13
Department of Mines, Haiti	44
Department of Mines, Malawi	65
Department of Mines, MOEA, Taiwan, Republic Of China	97
Department of Oceanography, Taiwan, Republic Of China	97
Department of Petroleum Affairs, Qatar	81
Department of Public Works (DOW), Aruba	3
Department of Public Works, Thailand	99
Department of Resources and Development, Micronesia, Federated States Of	69
Department of Science and Technology, India	48
Department of Seismology, Zaire	133
Department of Survey and Physical Planning, Cook Islands	31
Department of Surveying, Libyan Arab Jamahiriya	63
Department of Surveys and Lands, Botswana	13
Department of Surveys, Malawi	65
Department of Surveys, Swaziland	94
Department of the Surveyor General, Zimbabwe	134
Department of Water Affairs, Botswana	13
Department of Water Affairs, South Africa	90
Department of Water Resources, Gambia	41
Department of Water Resources, Taiwan, Republic Of China	97
Departmento Geografico Militar, Mexico	68

ORGANIZATION INDEX

Departmento Nacional de Aguas e Energia Eletrica (DNAEE), Brazil	14
Dept and Graduate Inst of Geology, China	27
Dept of Primary Industries and Energy, Australia	5
Desert Research Institute, Egypt	36
Dienst van het Kadaster (Office of the Land Registry), Netherlands Antilles	72
Dinamation International Society, United States	113
Direccao de Servicos de Geologia de Angola, Angola	2
Direccao-Geral dos Recursos Naturais, Portugal	81
Direccion de Geologia y Minas, Costa Rica	32
Direccion de Inversiones Mineras, Argentina	2
Direccion General de Hidrocarburos, Ecuador	35
Direccion General de Hidrocarburos, Panama	76
Direccion General de Minas e Hidrocarburos, Honduras	45
Direccion General de Mineria, Dominican Republic	35
Direccion General de Recursos Minerales, Panama	77
Direccion General del Instituto Geografico Nacional, Spain	92
Direccion General Sectorial de Minas y Geologia, Venezuela	130
Direccion Nacional de Mineria y Geologia (DINAMIGE), Uruguay	130
Dirección Nacional de Topografia, Uruguay	130
Direction de l'Hydraulique, Benin	12
Direction de la Cartographie et de la Topographie (DCT), Mauritania	66
Direction de la Pedologie, Togo	100
Direction de la Protection de la Nature (DPN), Mauritania	66
Direction de la Recherche Géologique et Minière, Niger	74
Direction de la Topographique, Benin	12
Direction des Hydrocarbures, Cote D'ivoire	32
Direction des Mines et de la Geologie, Algeria	1
Direction des Mines et de la Geologie, Cote D'ivoire	32
Direction du Service Topographique et du Cadastre, Niger	74
Direction for Urban and Rural Hydraulic, Senegal	88
Direction Generale de Mines et de la Geologie, Gabon	41
Direction Generale des Mines, de la Geologie et du Bureau National de Recherches Minieres, Togo	100
Direction Nationale des Mines et de la Geologie, Mali	66
Director of Building and Survey Department, Kuwait	62
Directorate General of Geology and Mineral Resources (DGGMR), Indonesia	49
Directorate General of Mines, Indonesia	50
Directorate General of Public Works, Lebanon	62
Directorate of Geological and Mining Research (D.R.G.M.), Chad	22
Directorate of Geological Survey, France	40
Directorate of Hydraulics and Water Clean Up (DHA), Chad	22
Directorate of Intelligence, United Arab Emirates	103
Directorate of Land Records and Survey, Bangladesh	10
Directorate of Military Survey, United Kingdom	104
Directorate of Mineral Resources (DMR), Indonesia	50
Directorate of National Mapping, Malaysia	65
Directorate of Petroleum, New and Renewable Energy (DPENR), Chad	22
Directorate, Surveys and Mapping, South Africa	90
Diretoria do Servico Geográfico do Exército (DSG), Brazil	14
Division Cartographie, Rwanda	86
Division de Recursos Geotermicos Del Ine, Nicaragua	73
Division of Earth Sciences, France	40

ORGANIZATION INDEX

Division of Land Improvement Planning and Exploration and Land/Water Resources, Greece	43
Division of Water, Environment and Forestry, South Africa	90
Donetsk Technical University, Ukraine	102
Doshisha University, Japan	55
Dublin Institute for Advanced Studies, Ireland	52
Earth Science Society of Libya, Libyan Arab Jamahiriya	63
Earth-Sciences Research Administration, Israel	52
Earthquake Engineering Research Institute, United States	113
Earthquake Research Department, Turkey	101
Earthquake Research Institute, Japan	55
East Midlands Geological Society, United Kingdom	104
East Tennessee Geological Society, United States	113
Ecole Nationale de Sciences Geodesiques, Algeria	1
Economic Development and Tourism, Canada	19
Economic Geology Publishing Company, United States	113
Economic Geology Research Unit, Australia	6
Ecosystem and Environmental Resources Directorate, Canada	19
Ecuadorian Geological and Geophysical Society, Ecuador	35
Edinburgh Geological Society, United Kingdom	104
Egypt Petroleum Exploration Society, Egypt	36
Egyptian Petroleum Research Institute, Egypt	36
Egyptian Remote Sensing Center, Egypt	36
Ehime University, Japan	55
El Paso Geological Society, United States	113
Empresa Colombiana de Minas Ecominas, Colombia	31
Empresa Minera del Centro del Peru (CENTROMIN PERU), Peru	77
Empresa Minera del Peru (MINPERO PERU), Peru	78
Empresa Nacional ADARO de Investigaciones Mineras, S.A., Spain	92
Empresa Nacional del Petróleo (ENAP), Chile	23
ENERGOPROEKT, Bulgaria	15
Energy and Natural Resources Division, Barbados	10
Energy Superintendency, El Salvador	37
Environment Division, Barbados	10
Environment Survey, Rwanda	86
Environmental and Engineering Geophysical Society, United States	113
Environmental Management Bureau, Philippines	79
Environmental National Committee, Guatemala	43
Environmental Research Institute of Michigan, United States	114
Environmental Service, Cyprus	33
Eötvös Loránd Geophysical Institute of Hungary, Hungary	46
Eötvös Loránd University, Hungary	46
Ethiopian Institute of Geological Survey (EIGS), Ethiopia	37
Ethiopian Mineral Resources Development Corp., Ethiopia	38
European Association of Geochemistry, United Kingdom	104
European Association of Geoscientists and Engineers, Netherlands	71
European Association of Science Editors, United Kingdom	104
European Geophysical Society, Germany	42
European Seismological Commission, Italy	53
European Union of Geosciences, United Kingdom	104
Exchange Centre for Scientific Literature, Finland	38

ORGANIZATION INDEX

Faculte des Sciences de Tunis, Tunisia 101
Faculty of Sciences, Nigeria 74
Far East Geological Institute (DVGI), Russian Federation 83
Far East Institute of Raw materials, Russian Federation 83
Far East State Technical University, Russian Federation 83
Federal Administration for International Scientific, Educational, Cultural and Technical Cooperation (FAISECTC), Yugosla 132
Federal Authority for Standardization and Geodesy, Austria 9
Federal Division for Water Protection, Switzerland 95
Federal Environment Agency, Austria 9
Federal Hydrometeorological Institute, Yugoslavia 132
Federal Institute for Testing and Research/Geotechnical Institute, Austria 9
Federal Office and Research Institute of Agriculture, Austria 9
Federal Office for Topographical Survey, Switzerland 95
Federal Office for Water Economy, Switzerland 95
Federation of Australian Scientific and Technological Societies, Australia 6
Field Conference of Pennsylvania Geologists, United States 114
Finnish Environment Institute, Finland 38, 39
Flinders University of SA, Australia 6
Florida Geological Survey, United States 114
Florida Paleontological Society Inc., United States 114
Four Corners Geological Society, United States 114
French Clay Group, France 40
French Palaeontological Association, France 40
Friends of Mineralogy, United States 114
Friends of Sherlock Holmes, United States 114
Friendship of Peoples Russian University, Russian Federation 83
Fuel Research Centre - PCSIR, Pakistan 76
Fukui University, Japan 55
Fukuoka University of Education, Japan 55
Fundação Instituto Brasileiro de Geografia e Estatistica (IBGE), Brazil 14
Fundacion Venezolana de Investigaciones Seismologicas (FUNVISIS), Venezuela 131
GDMB Society for Mining, Metallurgy, Resource and Environmental Technology, Germany 42
General Bureau of Aerophotography, Peru 78
General Bureau of Geology and Exploration, CNNC, China 28
General Command of Mapping, Turkey 101
General Direction of Studies and Major Hydraulic Works, Tunisia 101
General Directorate of Irrigation and Drainage, El Salvador 37
General Directorate of Mineral Research and Exploration (MTA), Turkey 101
General Directorate of Pollution Prevention and Control, Turkey 102
General Establishment of Geology and Mineral Resources, Syrian Arab Republic 96
General Petroleum and Mineral Organization (PETROMIN), Saudi Arabia 87
Geo Centrum Brabant, Netherlands 71
Geochemical Society of Japan, Japan 55
Geochemical Society, United States 114
Geodesy and Cartography Bureau, Hungary 46
Geodetic and Cartographic Administration, Czech Republic 34
Geodetic and Geophysical Surveys, India 48
Geodetic Institute, Czech Republic 34
Geodetic Institute, Denmark 34
Geodetic Institute, Finland 39

ORGANIZATION INDEX

Geographic Institute, Bulgaria	15
Geographical Institute of Burundi (IGEBU), Burundi	16
Geographical Survey Institute, Japan	55
Geoinstitut, Yugoslavia	132
Geological and Mineral Resources, Sudan	93
Geological and Mining Institute, Portugal	81
Geological and Mining Research Center, Zaire	133
Geological and Mining Research Office, Nigeria	74
Geological Association of Canada, Canada	20
Geological Association of New Jersey, United States	114
Geological Association, Germany	42
Geological Bureau of Ministry of Metallurgical Industry, China	28
Geological Department, Libyan Arab Jamahiriya	63
Geological Institute AS CR, Czech Republic	34
Geological Institute of the Russian Academy of Sciences (GIN RAS), Russian Federation	83
Geological Institute of the Slovak Academy of Sciences, Slovakia	89
Geological Institute of Uruguay, Uruguay	130
Geological Research and Mining Department, Libyan Arab Jamahiriya	63
Geological Research Institute (VSEGEI), Russian Federation	83
Geological Society of Africa, Kenya	60
Geological Society of Africa, Sudan	93
Geological Society of America, United States	114
Geological Society of Australia, Australia	6
Geological Society of Bulgaria, Bulgaria	15
Geological Society of Chile, Chile	23
Geological Society of Finland, Finland	39
Geological Society of France, France	40
Geological Society of Greece, Greece	43
Geological Society of Hong Kong, Hong Kong	45
Geological Society of India, India	48
Geological Society of Iowa, United States	115
Geological Society of Jamaica, Jamaica	54
Geological Society of Japan, Japan	55
Geological Society of Kentucky, United States	115
Geological Society of Maine, United States	115
Geological Society of Malaysia, Malaysia	65
Geological Society of Namibia, Namibia	70
Geological Society of New Zealand Inc., New Zealand	72
Geological Society of Peru, Peru	78
Geological Society of Poland, Poland	80
Geological Society of Portugal, Portugal	81
Geological Society of South Africa, South Africa	90
Geological Society of Spain, Spain	92
Geological Society of Sri Lanka, Sri Lanka	93
Geological Society of Sweden, Sweden	94
Geological Society of the North, France	40
Geological Society of the Oregon Country, United States	115
Geological Society of the Philippines, Philippines	79
Geological Society of Trinidad and Tobago, Trinidad And Tobago	100
Geological Society of Washington, D.C., United States	115
Geological Society, United States	114
Geological Survey and Mines Bureau, Sri Lanka	93

ORGANIZATION INDEX

Geological Survey and Mines Department, Swaziland 94
Geological Survey and Mines Department, Uganda 102
Geological Survey Department of Nigeria, Nigeria 74
Geological Survey Department, Botswana 13
Geological Survey Department, Cyprus 33
Geological Survey Department, Malawi 65
Geological Survey Department, Somalia 89
Geological Survey Department, Sri Lanka 93
Geological Survey Department, Zambia 133
Geological Survey Division, Australia 6
Geological Survey Division, Sierra Leone 88
Geological Survey of Alabama, United States 115
Geological Survey of Austria, Austria 9
Geological Survey of Bangladesh (GSB), Bangladesh 10
Geological Survey of Belgium, Belgium 11
Geological Survey of Denmark and Greenland, Denmark 34
Geological Survey of Estonia, Estonia 37
Geological Survey of Finland, Finland 39
Geological Survey of Ghana, Ghana 42
Geological Survey of Greenland, Greenland 43
Geological Survey of Hong Kong, Hong Kong 45
Geological Survey of India, India 48
Geological Survey of Iran, Iran 50
Geological Survey of Ireland, Ireland 52
Geological Survey of Israel, Israel 52
Geological Survey of Italy, Italy 53
Geological Survey of Japan (GSJ), Japan 56
Geological Survey of Kenya, Kenya 60
Geological Survey of Lithuania, Lithuania 64
Geological Survey of Lower Saxony, Germany 42
Geological Survey of Malaysia, Malaysia 65
Geological Survey of Morocco, Morocco 69
Geological Survey of Newfoundland and Labrador, Canada 20
Geological Survey of Norway, Norway 75
Geological Survey of Pakistan (GSP), Pakistan 76
Geological Survey of Rwanda, Rwanda 86
Geological Survey of Slovak Republic, Slovakia 89
Geological Survey of Sweden, Sweden 94
Geological Survey of the Netherlands, Netherlands 71
Geological Survey of Victoria, Australia 6
Geological Survey of Viet Nam, Viet Nam 131
Geological Survey of Western Australia, Australia 6
Geological Survey, Namibia 70
Geological Survey, Papua New Guinea 77
Geological Survey, South Africa 90
Geological, Mining and Metallurgical Society of India, India 48
Geologisches Institut, Switzerland 95
Geologist of Washington, D.C., United States 115
Geologists' Association, United Kingdom 105
Geology and Mineral Institute of Korea, Korea, Republic Of 61
Geology and Mining Service of Suriname, Suriname 94
Geology and Petroleum Office, Belize 11

ORGANIZATION INDEX

Magnetic Observatory, South Africa	90
Geophysical Exploration Co., Hungary	46
Geophysical Institute of Peru, Peru	78
Geophysical Institute, Bulgaria	15
Geophysical Observatory, Ethiopia	38
Geophysical Observatory, Nicaragua	73
Geophysical Society of Alaska, United States	115
Geophysical Society of Finland, Finland	39
Geophysical Society of Houston, United States	115
Geophysical Society of Peru, Peru	78
Geophysics Division, Iceland	47
Georgia Geologic Survey, United States	115
Georgia Geological Society, United States	115
Geoscience Information Society, United States	116
Geoscience Laboratory, Pakistan	76
Geoscience Society of Iceland, Iceland	47
Geotechnical Engineering International Resources Center, Thailand	99
Geotechnical Engineering Office, Hong Kong	45
Geothermal Energy Association, United States	116
Geothermal Resources Council, United States	116
Gerente División Exploración, Argentina	2
German Geological Society, Germany	42
German Geophysical Society, Germany	42
German Mineralogical Association, Germany	42
Gifu University, Japan	56
Global Systems Science Project, United States	116
Gold Bureau, Ministry of Metallurgical Industry, China	28
Gold Headquarters, Ministry of Metallurgical Industry, China	28
Graham Geological Society, United States	116
Grand Canyon National Park Research Library, United States	116
Grand Junction Geological Society, United States	116
Guadeloupe Volcano Observatory, Guadeloupe	43
Guatemalan Geological Society, Guatemala	44
Gubkin State Academy of Oil and Gas, Russian Federation	83
Gulf Coast Association of Geological Societies, United States	116
Gunma University, Japan	56
Guyana Geology and Mines Commission, Guyana	44
Hawaii Department of Land and Natural Resources, United States	116
Hebei College of Geology, China	28
Hebrew University of Jerusalem, Israel	53
Hellenic Army Geographic Service, Greece	43
Helsinki University of Technology, Finland	39
Heriot-Watt University, United Kingdom	105
High Commission for Scientific and Technological Research, Central African Republic	22
Himeji Institute of Technology, Japan	56
Hirosaki University, Japan	56
Hiroshima University, Japan	56
History of the Earth Sciences Society, United States	116
Hokkaido University of Education, Japan	56
Hokkaido University, Japan	56
Hong Kong Polytechnic University, Hong Kong	45
Housing Development Board, Singapore	88

147

ORGANIZATION INDEX

Houston Geological Society, United States	116
Hull Geological Society, United Kingdom	105
Humboldt Earthquake Education Center, United States	116
Hunan Bureau of Geology and Mineral Exploration and Development, China	28
Hungarian Geological Institute, Hungary	46
Hungarian Geological Society, Hungary	46
Hungarian Geological Survey, Hungary	46
Hungarian Hydrocarbon Institute, Hungary	46
Hungarian Hydrological Society, Hungary	46
Hydraulics Research Institute, Egypt	36
Hydro-Meteorological Bureau, Korea, Democratic People's Republic Of	61
Hydrocarbon Development Institute of Pakistan (HDIP), Pakistan	76
Hydrogeological Department, Somalia	89
Hydrologic Research Division, Suriname	94
Hydrological Branch, Ghana	42
Hydrological Branch, Zimbabwe	134
Hydrological Section, Tanzania, United Republic Of	98
Hydrological Service, Israel	53
Hydrology Department, Somalia	89
Hydrology Division, Sudan	93
Hydrology Section, Malta	66
Hydrology Section, Sri Lanka	93
Hydrometeorological Station, Guyana	44
Hydrometric Service, Ireland	52
Hyogo Kyoiku University, Japan	56
I.P.N. - E.S.I.A., Mexico	68
Ibaraki University, Japan	56
Iceland Geodetic Survey, Iceland	47
Iceland Glaciological Society, Iceland	47
Iceland National Energy Authority, Iceland	47
Icelandic Institute of Natural History, Iceland	47
Idaho Association of Professional Geologists, United States	116
Idaho Geological Survey, United States	116
Illinois Geological Society, United States	116
Illinois Groundwater Association, United States	116
Illinois State Geological Survey, United States	117
Imst. Sup. d'Estudes et Recherches Scientifique et Tech., Djibouti	34
Indian Association of Geohydrologists, India	48
Indian Bureau of Mines, India	48
Indian Geologists' Association, India	48
Indian Institute of Remote Sensing, India	48
Indian Petroleum Publishers, India	48
Indiana Geological Survey, United States	117
Indonesian Coal and Mining Association, Indonesia	50
Indonesian National Aeronautics and Space Institute, Indonesia	50
Industrial Research Institute, Ghana	43
Inland Geological Society, United States	117
Institut d'Hydraulique, Algeria	1
Institut d'Hydrotechnique et de Bonification, Algeria	1
Institut de Biologie et des Sciences de la Terre, Algeria	1
Institut des GeoSciences, Algeria	1
Institut des Sciences de la Terre, Algeria	1

ORGANIZATION INDEX

Institut Français de Recherche Scientifique por le Développement en coopération (ORSTOM), New Caledonia	72
Institut Français du Pétrole (IFP), France	40
Institut Géographique et Hydrographique National, Madagascar	64
Institut Géographique National, Belgium	11
Institut Geographique National, Central African Republic	22
Institut Géographique National, France	40
Institut Geographique, Congo	31
Institut Geotechnique de l'Etat, Belgium	11
Institut National de Cartographie, Algeria	1
Institut National de Topographie, Mali	66
Institute for Cartography, Yugoslavia	132
Institute for Development of Water Resources, Yugoslavia	132
Institute for Dynamics of the Geosphere, Russian Federation	83
Institute for Geological and Geophysical Research, Yugoslavia	132
Institute for Geological and Mineral Research (IRGM), Cameroon	16
Institute for Geological, Geophysical and Mining Exploration of Nuclear and other Mineral Resources, Yugoslavia	132
Institute for Geophysics, United States	117
Institute for Hydrogeology and Hydrotechnology, Yugoslavia	132
Institute for Mineralogy, Germany	42
Institute for Petroleum Research and Geophysics, Israel	53
Institute of Agronomic Research, Cameroon	16
Institute of Applied Geology, Taiwan, Republic Of China	97
Institute of Applied Geophysics, Taiwan, Republic Of China	97
Institute of Applied Science and Technology, Guyana	44
Institute of Earth Sciences, Academia Sinica, Taiwan, Republic Of China	97
Institute of Experimental Mineralogy (IEM), Russian Federation	83
Institute of Geologic Sciences, Ukraine	102
Institute of Geological and Nuclear Sciences Limited, New Zealand	73
Institute of Geological Sciences, Poland	80
Institute of Geology and Geochemistry, Russian Federation	84
Institute of Geology and Geophysics (IGG), Russian Federation	84
Institute of Geology and Mineral Exploration (IGME), Greece	43
Institute of Geology of Ore Deposits, Petrography, Mineralogy, Geochemistry (IGEM), Russian Federation	84
Institute of Geology, Armenia	3
Institute of Geology, China	28
Institute of Geology, Korea, Democratic People's Republic Of	61
Institute of Geology, Taiwan, Republic Of China	97
Institute of Geophysics, Poland	80
Institute of Geophysics, Viet Nam	131
Institute of Geosciences, Viet Nam	131
Institute of Hydrogeology and Engineering Geology, China	28
Institute of Hydrology, United Kingdom	105
Institute of Lithosphere (ILSAN), Russian Federation	84
Institute of Marine Affairs, Trinidad And Tobago	100
Institute of Marine Geology and Chemistry, Taiwan, Republic Of China	97
Institute of Marine Geology and Geophysics (IMGIG), Russian Federation	84
Institute of Meteorology and Water Management, Poland	80
Institute of Minerals, Energy and Construction, Australia	6
Institute of Mining Research, Zimbabwe	134

ORGANIZATION INDEX

Institute of Oil and Gas Research, Viet Nam 131
Institute of Paleobiology, Poland 80
Institute of Petroleum Exploration, India 48
Institute of Physics of the Earth (IFZ), Russian Federation 84
Institute of Rock and Mineral Analysis, China 29
Institute of Seismology, Nicaragua 73
Institute of Seismology, Volcanology, Meteorology and Hydrology (INSIVUMEH), Guatemala 44
Institute of Tectonics and Geophysics (ITIG), Russian Federation 84
Institute of the Earth's Crust (IZK), Russian Federation 84
Institute of the Expanding Earth, Jamaica 54
Institute of Volcanology (IV), Russian Federation 84
Institute of Water Problems, Bulgaria 15
Instituti Hidrometeorologjik, Albania 1
Institution of Mining Engineers, United Kingdom 105
Instituto Andaluz de Ciencias de la Terra, Spain 92
Instituto Costarricense de Electricidad (ICE), Costa Rica 32
Instituto Cubano de Geodesia y Cartografia, Cuba 33
Instituto Cubano de Hidrografia, Cuba 33
Instituto de Ciencias de la Tierra "Jaime Almera", Spain 92
Instituto de Geocronologia y Geologia Isotopica, Argentina 2
Instituto de Geofisica y Astronomia, Cuba 33
Instituto de Geografia, Cuba 33
Instituto de Geologia y Paleontologia, Cuba 33
Instituto de Geologia, Cuba 33
Instituto de Geologia, Mexico 68
Instituto de Geologica Economica, Spain 92
Instituto de Hidraulica e Hidrologia, Bolivia 12
Instituto de Hidrología, Meteorología, y Estudios Ambientales, Colombia 31
Instituto de Investigacion de Recuros Naturales (IREN), Chile 23
Instituto de Investigaciones Electicas, Mexico 68
Instituto de Investigaciones en Geosciencias, Mineria y Quimica (INGEOMINAS), Colombia 31
Instituto de Oceanologia, Cuba 33
Instituto de Recursos Hidráulicos y Electrificación, Panama 77
Instituto Ecuatoriano de Mineria (INEMIN), Ecuador 35
Instituto Geofisico, Ecuador 36
Instituto Geografico "Agustin Codazzi", Colombia 31
Instituto Geográfico Militar (IGM), Ecuador 36
Instituto Geografico Militar, Argentina 2
Instituto Geografico Militar, Bolivia 12
Instituto Geográfico Militar, Chile 23
Instituto Geografico Nacional "Tommy Guardia", Panama 77
Instituto Geografico Nacional (IGN), Honduras 45
Instituto Geografico Nacional, Costa Rica 32
Instituto Geografico Universitario, Dominican Republic 35
Instituto Geologico Minero y Metalugico (Geological Mining and Metallurgical Institute) INGEMMET, Peru 78
Instituto Mexicano de Tecnologia del Agua, Mexico 68
Instituto Mexicano del Petroleo, Mexico 68
Instituto Minero Metalurgico (IMM), Bolivia 12
Instituto Nacional de Acueductos y Alcantarillados (IDAAN), Panama 77

ORGANIZATION INDEX

Instituto Nacional de Ciencia y Tecnica Hidricas, Argentina	2
Instituto Nacional de Investigacíon Nuclear, Mexico	68
Instituto Nacional de Meteorologia e Hidrologia (INAMHI), Ecuador	36
Instituto Nacional de Pesca (INAPE), Uruguay	130
Instituto Nacional de Recursos Hidraulicos, Dominican Republic	35
Instituto Nacional de Recursos Minerales (ICRM), Cuba	33
Instituto Nacional de Recursos Nacionales Renovables, Panama	77
Instituto Nicaraguense de Estudios Territoriales, Nicaragua	73
Instituto Nicaraguense de Minas e Hidrocarburos (INMINE), Nicaragua	73
Instituto Nicaraguense de Recursos Naturales y del Ambiente (IRENA), Nicaragua	74
Instituto Oceanografico de la Armada (INOCAR), Ecuador	36
Instituto Português de Cartografia e Cadastro, Portugal	81
Instituto Tecnologico de Cd. Madero, Mexico	68
Instituto Tecnologico Geominero de Espana, Spain	92
Institutul de Geografie, Romania	82
Institutul de Meteorologie si Hidrologie, Romania	82
Institutul Geologic al Romaniei, Romania	82
Instytut Geodezji i Kartografii, Poland	80
Instytut Geografii, Poland	80
Instytut Geologiczny, Poland	80
Inter-Union Commission on the Lithosphere, Canada	20
Interafrican Committee for Water Resources Studies, Burkina Faso	16
Intergovernmental Oceanographic Commission, France	40
International Association for Great Lakes Research, United States	117
International Association for Mathematical Geology, United States	117
International Association for the Physical Sciences of the Ocean, United States	117
International Association for the Study of Clays, Belgium	11
International Association of Engineering Geology, France	40
International Association of Geochemistry and Cosmochemistry, Canada	20
International Association of Geochemistry and Cosmochemistry, United States	117
International Association of Geodesy, Denmark	34
International Association of Geomagnetism and Aeronomy, United Kingdom	105
International Association of Hydrogeologists, Canada	20
International Association of Hydrogeologists, Netherlands	71
International Association of Hydrogeologists, United Kingdom	105
International Association of Hydrogeologists/U.S. National Chapter, United States	117
International Association of Hydrological Sciences, Canada	20
International Association of Sedimentologists, Switzerland	95
International Association of Seismology and Physics of the Earth's Interior, India	49
International Association of Volcanology and the Chemistry of the Earth's Interior (IAVCEI), Australia	6
International Association on the Genesis of Ore Deposits, Canada	20
International Association on the Genesis of Ore Deposits, Czech Republic	34
International Center for Arid and Semiarid Land Studies, United States	117
International Center for Training and Exchanges in the Geosciences, France	40
International Council for the Exploration of the Sea, Denmark	34
International Federation of Palynological Societies, United States	117
International Geological Correlation Programme (IGCP), France	40
International Geosphere-Biosphere Programme, Sweden	95
International Glaciological Society, United Kingdom	105
International Institute for Geothermal Research, Italy	53
International Institute of Aerospace Survey and Earth Sciences, Netherlands	71

ORGANIZATION INDEX

International Institute of Earthquake Prediction Theory and Mathematical Geophysics,
Russian Federation ... 84
International Institute of Volcanology, Italy ... 54
International Landslide Research Group, United States ... 117
International Marine Minerals Society, United States ... 117
International Mineralogical Association, Germany ... 42
International Paleontological Association, Japan ... 56
International Peat Society, Finland ... 39
International Permafrost Association, United States ... 118
International Society of Soil Science, Austria ... 9
International Society of Soil Science, Netherlands ... 71
International TOGA Project Office, Switzerland ... 95
International Tsunami Information Center, United States ... 118
International Union for Quaternary Research, Norway ... 75
International Union for Quaternary Research, Switzerland ... 95
International Union for Quaternary Research, United States ... 118
International Union of Crystallography, United Kingdom ... 105
International Union of Geodesy and Geophysics, France ... 41
International Union of Geological Sciences (IUGS), Norway ... 75
International Water Supply Association, United Kingdom ... 105
Iowa Department of Natural Resources, United States ... 118
Iraqi Geologists' Union, Iraq ... 51
IRIS Data Management Center, United States ... 118
IRIS, United States ... 118
Irish Association for Economic Geology, Ireland ... 52
Irish Geological Association, Ireland ... 52
Irkutsk State Technical University, Russian Federation ... 84
Irkutsk State University, Russian Federation ... 84
Irrigation and Water Resources Department, Syrian Arab Republic ... 96
Irrigation Department, Myanmar ... 70
Israel Geological Society, Israel ... 53
Italian Geological Society, Italy ... 54
Italian Geotechnical Society, Italy ... 54
Italian Glaciological Committee, Italy ... 54
Italian Group of A.I.P.E.A., Italy ... 54
Italian Military Geographic Institute, Italy ... 54
Italian Paleontological Society, Italy ... 54
Ivanovo-Frankivsk Institute for Oil & Gas, Ukraine ... 102
Iwate University, Japan ... 56
Jahangirnagar University, Bangladesh ... 10
James Cook University of North Queensland, Australia ... 6
Japanese Association of Mineralogists, Petrologists and Economic Geologists, Japan ... 56
Japanese National Committee of Geology, Japan ... 56
Joint Association of Geophysics, United Kingdom ... 105
Joint Oceanographic Institute, United States ... 118
JSC Vilnius Hydrogeology Ltd., Lithuania ... 64
Jurong Town Corporation, Singapore ... 88
Kagawa University, Japan ... 56
Kagoshima University, Japan ... 56
Kanazawa University, Japan ... 57
Kansas Geological Society and Library, United States ... 118
Kansas Geological Survey, United States ... 118

ORGANIZATION INDEX

Kazan State University, Russian Federation 84
Kentucky Geological Survey, United States 118
Key Centre for Teaching and Research in Strategic Mineral Deposits, Australia 6
Khabarovsk State Technical University, Russian Federation 84
Kharkov State University, Ukraine 102
Khon Kaen University, Thailand 99
Kiev State University, Ukraine 102
King Fahd University of Petroleum and Minerals, Saudi Arabia 87
Kobe University, Japan 57
Kochi University, Japan 57
Korea Institute of Geology, Mining and Materials (KIGAM), Korea, Republic Of 61
Korea Ocean Development Research Institute (KORDI), Korea, Republic Of 61
Krasnoyarsk Institute of Non-ferrous Metals, Russian Federation 84
Krivoy Rog Technical University, Ukraine 102
Kuban State University, Russian Federation 84
Kumamoto University, Japan 57
Kuwait Institute for Scientific Research, Kuwait 62
Kyoto University, Japan 57
Kyushu University, Japan 57
Laboratorio Nacional de Engenharia Civil, Portugal 81
Laboratory for Geotechnology of Weak Earth, Bulgaria 15
Laboratory of Seismic Mechanics and Earthquake Engineering, Bulgaria 15
Lafayette Geological Society, United States 118
Lamont-Doherty Earth Observatory, United States 118
Land Improvement Works Division (D7), Greece 43
Land Registry Service, Netherlands 71
Landeshydrologie und Geologie, Switzerland 95
Lands and Survey Department, Fiji 38
Lands and Survey Department, Samoa 87
Lands and Surveys Department, Guyana 44
Latrobe University College of Advanced Education, Australia 7
Latrobe University, Australia 7
Liberian Cartographic Service, Liberia 63
Liberian Geological Survey, Liberia 63
Liberian Hydrological Service, Liberia 63
Limnological Institute, Russian Federation 84
Liverpool Geological Society, United Kingdom 106
Lomonosov Moscow State University, Russian Federation 85
Louisiana Geological Survey, United States 119
Lunar and Planetary Institute, United States 119
Lviv State University, Ukraine 102
Macquarie University, Australia 7
Main Computer Center (GLAVNIVTZ), Russian Federation 85
Maine Geological Survey, United States 119
Manchester Geological Association, United Kingdom 106
Manitoba Energy and Mines, Canada 20
Marine Geological Institute, Indonesia 50
Marine Geology Group, Spain 92
Marine Research Institue, Iceland 47
Marshall Islands Marine Resources Authority (MIMRA), Marshall Islands 66
Maryland Geological Survey, United States 119
Massachusetts Office of Environmental Affairs, United States 119

ORGANIZATION INDEX

Massey University, New Zealand	73
Mauritius Sugar Industry Research Institute, Mauritius	67
Meijo University, Japan	57
Mekong River Commission Secretariat, Thailand	99
Meteoritical Society, United Kingdom	106
Meteorological and Geophysical Agency, Indonesia	50
Meteorological Department, Thailand	99
Meteorology and Hydrology Department, Myanmar	70
Mexican Association of Exploration Geophysicists, Mexico	68
Mexican Association of Petroleum Geologists, Mexico	68
Mexican Geological Society, Mexico	68
Mexican Geophysical Union, Mexico	68
Michigan Basin Geological Society, United States	119
Michigan Earth Science Teachers Association, United States	119
Michigan Geological Survey Division, United States	119
Mie University, Japan	57
Military Geodetic Service, Paraguay	77
Military Geographic Institute, Guatemala	44
Military Geographical Institute, Czech Republic	34
Mineral Exploration Board, Yemen	132
Mineral Research and Exploration Institute, Turkey	102
Mineral Resources Department, Fiji	38
Mineral Resources Exploration Company, Brazil	14
Mineral Resources Tasmania, Australia	7
Mineralogical Association of Canada, Canada	20
Mineralogical Society of America, United States	119
Mineralogical Society of Japan, Japan	57
Mineralogical Society, United Kingdom	106
Minerals Management Service, United States	119
Mines and Geological Department, Kenya	61
Mines and Geosciences Bureau, Philippines	79
Mines Division, Hong Kong	45
Mines Service, Cyprus	33
Mining Association of British Columbia, Canada	20
Mining Association, Viet Nam	131
Mining Industry and Technical Services, Zambia	133
Ministère de L'Agriculture et de la Mise en Valeur Agricole, Morocco	69
Ministere de l'Equipement, des Mines et de l'Energie, Togo	100
Ministere de l'Hydraulique, Algeria	1
Ministere de l'Industrie, de l'Energie et de Mines (MIEM), Madagascar	64
Ministere des Affaires Economiques, Belgium	11
Ministere des Mines et de la Geologie, Mauritania	66
Ministere des Mines, Cote D'ivoire	33
Ministère des Travaux Publics, Morocco	69
Ministere des Travaux Publiques, Belgium	11
Ministere des Travaux Publiques, Djibouti	35
Ministere Equipement et Transp., Belgium	11
Ministerio de Agricultura y Ganaderia, El Salvador	37
Ministerio de Ciencia y Tecnologia, Costa Rica	32
Ministerio de Construccion y Transporte Instituto Nicaraguese de Estudios Territoriales (INETER), Nicaragua	74
Ministerio de Economia y Obras y Servicios Publicos, Argentina	3

ORGANIZATION INDEX

Ministerio de Industria y Energia, Spain	93
Ministerio de Recursos Naturales, Energia y Minas, Costa Rica	32
Ministrey of Public Works, El Salvador	37
Ministry of Agriculture and Cooperatives, Thailand	99
Ministry of Agriculture and Fisheries, Oman	75
Ministry of Agriculture and Fisheries, United Arab Emirates	103
Ministry of Agriculture and Forestry/Hydrographical Central Bureau, Austria	9
Ministry of Agriculture and Lands, Solomon Islands	89
Ministry of Agriculture and Mining, Jamaica	54
Ministry of Agriculture and Natural Resources, Gambia	41
Ministry of Agriculture and Water, Saudi Arabia	87
Ministry of Agriculture, El Salvador	37
Ministry of Agriculture, Guinea	44
Ministry of Agriculture, Trade, Industry and Tourism, Dominica	35
Ministry of Agriculture, Tunisia	101
Ministry of Agriculture, Water and Rural Development, Namibia	70
Ministry of Electricity and Water Desalinization, United Arab Emirates	103
Ministry of Electricity and Water, Kuwait	62
Ministry of Energy and Energy Industries, Trinidad And Tobago	100
Ministry of Energy and Mineral Resources, Bangladesh	10
Ministry of Energy and Mines, Burundi	16
Ministry of Energy and Mines, Guatemala	44
Ministry of Energy and Mines, Tunisia	101
Ministry of Energy, Mines, Geology and Hydraulics, Central African Republic	22
Ministry of Energy, Water and Mineral Resources, Solomon Islands	89
Ministry of Environmental Protection and Water Management, Hungary	46
Ministry of Geology and Mineral Resources, China	29
Ministry of Geology and Mineral Resources, Mongolia	69
Ministry of Higher Education, Computer Services and Scientific Research, Cameroon	16
Ministry of Housing, Lands, and Town and Country Planning, Mauritius	67
Ministry of Hydro-Electric Resources, Lebanon	62
Ministry of Hydrology and Environment, Nigeria	74
Ministry of Internal Affairs, Sudan	94
Ministry of Lands, Mines and Energy, Liberia	63
Ministry of Lands, Survey and Natural Resources, Tonga	100
Ministry of Local Government and Rural Development, Kiribati	61
Ministry of Metallurgical Industry, China	29
Ministry of Mineral Resources and Water Affairs, Botswana	13
Ministry of Mines and Energy (MME), Brazil	14
Ministry of Mines and Energy, Ethiopia	38
Ministry of Mines and Energy, Namibia	71
Ministry of Mines and Geology, Guinea	44
Ministry of Mines, Energy, Industry and Handicraft, Nigeria	74
Ministry of Natural Resources and Energy, Swaziland	94
Ministry of Natural Resources Development, Kiribati	61
Ministry of Natural Resources, Belize	11
Ministry of Natural Resources, Guinea-bissau	44
Ministry of Natural Resources, Solomon Islands	89
Ministry of Natural Resources, Uganda	102
Ministry of Oil and Mineral Resources, Yemen	132
Ministry of Petroleum and Mineral Resources, Saudi Arabia	87
Ministry of Petroleum and Mineral Resources, United Arab Emirates	103

ORGANIZATION INDEX

Ministry of Petroleum and Minerals, Oman 75
Ministry of Planning and Mobilisation, Trinidad And Tobago 100
Ministry of Public Works and Town Development, Central African Republic 22
Ministry of Research and Environmental Affairs, Malawi 65
Ministry of Rural Development, Central African Republic 22
Ministry of the Environment and Quality of Life, Mauritius 67
Ministry of the Environment, Singapore 88
Ministry of Transport and Public Works, Netherlands 72
Ministry of Water and Power, Iran 50
Ministry of Water Resources, China 29
Ministry of Water, Energy and Mining, Lesotho 62
Ministry of Water, Lands, Game Fishing, and Tourism, Central African Republic 22
Minnesota Geological Survey, United States 119
Mississippi Geological Society, United States 119
Mississippi Office of Geology, United States 119
Missouri Department of Natural Resources, United States 119
Miyagi University of Education, Japan 57
Miyazaki University, Japan 57
Monash University, Australia 7
Mongolian Geological and Mining Society, Mongolia 69
Montana Bureau of Mines and Geology, United States 119
Montana Geological Society, United States 120
Monterey Bay Geological Society, United States 120
Moscow Geological Prospecting Institute, Russian Federation 85
Moscow State Academy of Exploration Geology, Russian Federation 85
Moscow State Open University, Russian Federation 85
Muroran Institute of Technology, Japan 57
Muzeum Ziemi PAN, Poland 80
Myanmar Oil Corporation, Myanmar 70
Nacional de Metales Preciosos S.A., Colombia 31
Nagasaki University, Japan 57
Nagoya University, Japan 57
Nanjing University, China 29
Nara University of Education, Japan 57
National Aeronautics and Space Administration, United States 120
National Association of Black Geologists and Geophysicists, United States 120
National Association of Geoscience Teachers, United States 120
National Association of State Boards of Geology, United States 120
National Bureau of Surveying and Mapping, China 29
National Cartographic Center (NCC), Iran 51
National Center for Earthquake Engineering Research, United States 120
National Cheng Kung University, Taiwan, Republic Of China 97
National Commission of Geology, Spain 93
National Committee of Geography, Geodesy and Geophysics, Chile 23
National Crude Oil and Natural Gas Trust, Hungary 47
National Director of the Metal Mining Economy, Argentina 3
National Directorate for Geography and Mapping, Mozambique 69
National Directorate of Geology, Mozambique 69
National Directorate of Hydrology, Mozambique 70
National Earth Orientation Service, United States 120
National Earth Science Teachers Association, United States 120
National Environmental Protection Agency, China 29

ORGANIZATION INDEX

National Geodetic Survey, United States 120
National Geographic Department, Lao People's Democratic Republic 62
National Geographic Institute, El Salvador 37
National Geographic Institute, Peru 78
National Geographic Organization, Iran 51
National Geography Institute, Korea, Republic Of 62
National Geomatics Center of China (NGCC), China 29
National Geophysical Research Institute, India 49
National Ground Water Association, United States 120
National Ground Water Information Center, United States 120
National Institute for Earth Physics, Romania 82
National Institute for Environmental Studies, Japan 58
National Institute of Cartography, Cameroon 16
National Institute of Geological Sciences, Philippines 79
National Institute of Meteorology and Geophysics, Portugal 81
National Institute of Natural Resources (INRENA), Peru 78
National Institute of Oceanography (NIO), Pakistan 76
National Institute of Oceanography, India 49
National Iranian Oil Company, Iran 51
National Land Survey of Finland, Finland 39
National Mapping and Resource Information Authority (NAMRIA), Philippines 79
National Mining Association, United States 120
National Mining Hall of Fame and Museum, United States 120
National Mining Service, Argentina 3
National Mining Society, Peru 78
National Oceanic and Atmospheric Administration (NOAA), United States 121
National Office of Geology, Algeria 2
National Office of Mining Research, Nigeria 74
National Pastoral and Village Hydraulics Office (ONHPV), Chad 22
National Remote Sensing Agency, India 49
National Remote Sensing Center, Syrian Arab Republic 96
National Research Council of Thailand, Thailand 99
National Research Institute for Astronomy and Geophysics, Egypt 36
National Research Institute for Resources and Environment, Japan 58
National Research Institute of Geology of Foreign Countries (VZG), Russian Federation . . . 85
National Science Foundation, United States 121
National Service of Meteorology and Hydrology, Peru 78
National Society of Mining, Chile 23
National Speleological Society, United States 121
National Taiwan Normal University, Taiwan, Republic Of China 97
National Taiwan University, Taiwan, Republic Of China 97
National University of Malaysia, Malaysia 65
National Water Commission, Jamaica 55
National Water Research Institute, Canada 20
National Water Resources Board, Philippines 79
Natural Resources Agency, Guyana 44
Natural Resources Authority (NRA), Jordan 60
Natural Resources Canada, Canada 21
Naval Bureau of Hydrography and Navigation, Peru 78
Navy Hydrographic Institute, Italy 54
Nebraska Geological Society, United States 121
Nepal Geological Society, Nepal 71

ORGANIZATION INDEX

Nevada Bureau of Mines and Geology, United States 121
New Brunswick Minerals and Energy Division, Canada 21
New England Intercollegiate Geological Conference, United States 121
New England States Emergency Consortium (NESEC), United States 121
New Hampshire Geological Survey, United States 121
New Jersey Geological Survey, United States 121
New Mexico Bureau of Mines and Mineral Resources, United States 121
New Mexico Geological Society Inc., United States 121
New Orleans Geological Society, United States 121
New York State Geological Association, United States 122
New York State Geological Survey, United States 122
New Zealand Antarctic Programme, New Zealand 73
New Zealand Hydrological Society, New Zealand 73
Nihon University, Japan 58
Niigata University, Japan 58
No. (1) Mining Corporation, Myanmar 70
No. (2) Mining Corporation, Myanmar 70
No. (3) Mining Corporation, Myanmar 70
Nordic Volcanological Institute, Iceland 47
North American Commission on Stratigraphic Nomenclature, Canada 21
North Carolina Geological Survey, United States 122
North Caucasus Mining and Metallurgical Institute, Russian Federation 85
North Dakota Geological Survey, United States 122
North Texas Geological Society, United States 122
Northeastern Interdisciplinary Scientific Research Institute, Russian Federation 85
Northeastern Science Foundation Inc., affiliated with Brooklyn College of the City University
of New York, United States 122
Northern California Geological Society, United States 122
Northern Illinois University Geophysical Society, United States 122
Northern Ohio Geological Society, United States 122
Northern Territory Geological Survey, Australia 7
Northwest Geological Society, United States 122
Northwest Mining Association, United States 122
Northwest Petroleum Association, United States 122
Northwest Territories Geology Division, Canada 21
Northwest University, China 30
Norwegian Geological Society, Norway 75
Norwegian Institute for Water Research, Norway 75
Norwegian Petroleum Society, Norway 75
Norwegian Polar Institute, Norway 75
Norwegian Water Resources and Energy Administration, Norway 75
Nova Scotia Department of Natural Resources, Canada 21
Novocherkassk State Technical University, Russian Federation 85
Novosibirsk State University, Russian Federation 85
Nuclear Regulatory Commission, United States 122
Obervatorio San Calixto, Bolivia 12
Observatoir Volcanologique du Piton de la Fournaise, Reunion 81
Observatoire Volcanologique de la Montagne Pelee, Martinique 66
Observatorio Meteorogico e Magnetico, Angola 2
Ocean Drilling Program, United States 122
Oceanographic & Limnological Research Institute, Israel 53
Oceanography Institute, Bulgaria 16

ORGANIZATION INDEX

Odessa State University, Ukraine	102
Office Béninois des Mines, Benin	12
Office Mauritanien de Recherches Geologiques (OMRG), Mauritania	67
Office of Hydrographic Affairs, Korea, Republic Of	62
Office of Mineral Resources, Paraguay	77
Office of Mines and Geology, Portugal	81
Office of Overseas Scientific and Technical Research, Martinique	66
Office of the Prime Minister, Cook Islands	31
Office of Topography and Cartography, Tunisia	101
Oficina Nacional del Catastro, Honduras	45
Ohio Division of Geological Survey, United States	122
Ohio Geological Society, United States	123
Oil and Gas Development Corporation (OGDC), Pakistan	76
Oil and Gas Institute, Poland	80
Oil and Gas Laboratories and Engineering Group, Hungary	47
Oil and Natural Gas Commission, India	49
Oita Univesity, Japan	58
Okayama University of Science, Japan	58
Okayama University, Japan	58
Okhta Industrial Institute, Russian Federation	85
Oklahoma City Geological Society Inc., United States	123
Oklahoma Geological Survey, United States	123
Ontario Geological Survey, Canada	21
Ontario Mining Association, Canada	21
Ordinance Survey, Ireland	52
Ordnance Survey International, United Kingdom	106
Oregon Department of Geology and Mineral Industries, United States	123
Organization for Complex Geological Research, Yugoslavia	133
Organization for the Development of the Gambia River Basin (OMVG), Senegal	88
ORSTOM Dakar, Senegal	88
Osaka City University, Japan	58
Osaka Institute of Technology, Japan	58
Osaka Kyoiku University (Ikeda Campus), Japan	58
Osaka Kyoiku University (Tennoji Campus), Japan	58
Osaka University, Japan	58
Oxford University, United Kingdom	106
Pacific Northwest National Laboratories, United States	123
Pacific Oceanological Institute, Far Eastern Branch, Russian Academy of Sciences (POI FEBRAS), Russian Federation	85
Pakistan Meteorological Department, Pakistan	76
Pakistan Mineral Development Corporation (PMDC), Pakistan	76
Pakistan Space and Upper Atmosphere Research Commission (SUPARCO), Pakistan	76
Palaeontographical Society, United Kingdom	107
Palaeontological Association, United Kingdom	107
Palaeontological Society of China, China	30
Palaeontological Society of Japan, Japan	58
Paleontological Research Institution, United States	123
Paleontological Society of Russian Academy of Sciences, Russian Federation	85
Paleontological Society, Germany	42
Paleontological Society, United States	123
Palynological Society of India, India	49
Pander Society, United States	123

ORGANIZATION INDEX

Panhandle Geological Society, United States	123
Para Cobre, Panama	77
Peak District Mines Historical Society, United Kingdom	107
Peking University, China	30
Pennsylvania Bureau of Topographic and Geologic Survey, United States	123
Perm State University, Russian Federation	85
Peruvian Institute of the Sea (IMARPE), Peru	78
Peruvian Mining and Petroleum Society, Peru	78
Petrobangla, Bangladesh	10
Petrol Ofisi A.S Genel Mudurlugu, Turkey	102
Petroleo Brasileiro S.A. (PETROBRAS), Brazil	14
Petroleum and Gas Engineering Department, Iraq	51
Petroleum Authority of Thailand, Thailand	99
Petroleum Exploration Society of Great Britain, United Kingdom	107
Petroleum Research Institute, Egypt	36
Petroleum Unit, Brunei Darussalam	14
Petroliam Nasional Berhad (PETRONAS), Malaysia	65
Philippine Atmospheric, Geophysical, and Astronomical Services Administration, Philippines	79
Philippine Institute of Volcanology and Seismology (PHILVOLCS), Philippines	79
Philippine Society of Mining, Metallurgical, and Geological Engineers, Philippines	79
Photogrammetric Society, United Kingdom	107
Pittsburgh Association of Petroleum Geologists, United States	123
Pittsburgh Geological Society, United States	123
Planning Division, Libyan Arab Jamahiriya	63
Plechanov St. Petersburg State Mining Institute, Russian Federation	85
Polar Continental Shelf Project, Canada	17
Portuguese Commission on International Hydrologic Program, Portugal	81
Prince Edward Island Department of Energy and Minerals, Canada	21
Projeto RADAM, Brazil	14
Public Authority for Water Resources, Oman	75
Public Petroleum Corporation (DEP), Greece	43
Public Works Department, Singapore	88
Public Works Research Institute, Japan	58
Puerto Rico Geological Survey Division, United States	123
Quaternary Research Association, Ireland	52
Quebec Ministre des Ressources Naturelles, Canada	21
Queens University, Ireland	52
Queensland University of Technology, Australia	7
Rabaul Volcano Observatory, Papua New Guinea	77
Rand Afrikaans University, South Africa	91
Real Sociedad Geografica, Spain	93
Regional Center for Services in Surveying, Mapping and Remote Sensing, Kenya	61
Comisia Romania pentru Activitati Spatiale, Romania	82
Remote Sensing Society, United Kingdom	107
Research Institute of Geology and Mineral Resources, Viet Nam	131
Research Institute of Mineral Resources, Kazakhstan	60
Research Institute of Petroleum, China	30
Resource Development Corporation (RDC), Pakistan	76
RHB New College, United Kingdom	107
Rhode Island State Geologist, United States	123
Rhodes University, South Africa	91

ORGANIZATION INDEX

RMI Environmental Protection Authority, Marshall Islands	66
Rocky Mountain Association of Geologists, United States	124
Rostov State University, Russian Federation	85
Roswell Geological Society Inc., United States	124
Royal Geographical Society (with The Institute of British Geographers), United Kingdom	107
Royal Geological and Mining Society of The Netherlands, Netherlands	72
Royal Irrigation Department, Thailand	100
Royal Jordanian Geographic Centre, Jordan	60
Royal Melbourne Institute of Technology, Australia	7
Royal Observatory, Hong Kong	45
Royal Scientific Society, Jordan	60
Royal Society, United Kingdom	107
Royal Spanish Society for Natural History, Spain	93
Royal Thai Survey Department, Thailand	100
Rural Water Administration, Sudan	94
Rural Water Supply Board, Swaziland	94
Russell Society, United Kingdom	107
Russian Academy of Sciences, Russian Federation	85
Russian Federal Service for Geodesy and Cartography (ROSGEOCART), Russian Federation	86
Russian State Committee on Geology (ROSKOMNEDRA), Russian Federation	86
Rÿks Universiteit Utrecht, Netherlands	72
Sacramento Petroleum Association, United States	124
Saga University, Japan	58
Saitama University, Japan	58
Saline Water Conservation Corporation (SWCC), Saudi Arabia	87
San Angelo Geological Society, United States	124
San Diego Society of Natural History, United States	124
San Diego State University Geology Alumni, United States	124
Sarhad Development Authority, Pakistan	76
Saskatchewan Energy and Mines, Canada	21
Saskatchewan Geological Society, Canada	21
Science Centre Board, Singapore	89
Scientific Industrial Enterprise on Super Deep Drilling (NEDRA), Russian Federation	86
Scientific Research Council, Jamaica	55
Secretaría de Agricultura, Ganadería y Desarrollo Rural, Mexico	68
Secretaria de Estado de Ciencia y Tecnologia, Argentina	3
Secretaria de Medio Ambiente, Mexico	68
Secretariat d'Etat aux Forets et a la Mise en Valeur des Terres, Algeria	2
Seismic Exploration Company, Iraq	51
Seismic Research Unit, Trinidad And Tobago	101
Seismic Service, Italy	54
Seismological Laboratory (MS 174), United States	124
Seismological Observatory of HAS, Hungary	47
Seismological Society of America, United States	124
Seismological Society of Japan, Japan	58
Seismology Unit, Iraq	51
Senegal Geographic Service, Senegal	88
SEPM (Society for Sedimentary Geology), United States	124
Service Autonoma de Geohrafica y Cartografia Nacional, Venezuela	131
Service de Geodesie et de Cartographie, Haiti	45
Service de l'Hydraulique, Djibouti	35

161

ORGANIZATION INDEX

Service de la Meteorologie et de l'Hydrologie, Luxembourg 64
Service des Mines et de l'Energie, New Caledonia 72
Service des Mines et de la Geologie, Congo 31
Service des Mines et de la Geologie, Madagascar 64
Service du Drainage et de l'Assainissement, Cote D'ivoire 33
Service Geographique de l'Armee, Syrian Arab Republic 96
Service Geographique Nationale, Viet Nam 131
Service Geologique de Belgique, Belgium 11
Service Geologique, Luxembourg 64
Service Hydrologique de Genie Rural, Niger 74
Service Hydrologique, Mali 66
Service Hydrometeorologique, Viet Nam 132
Service Meteorologique et Hydrologique, Haiti 45
Service of Meteorology and Hydrology, Lao People's Democratic Republic 62
Service Topographique et du Cadastre, Congo 31
Service Topographique, New Caledonia 72
Services des Travaux Publics, Reunion 82
Servicio Aerofotogrametrico, Chile 23
Servicio Autonomo Nacional de Acueductos y Alcantarillados (SANAA), Honduras 45
Servicio de Geologia, Spain 93
Servicio de Hidrografia Naval, Argentina 3
Servicio de Oceanografia, Hidrografia y Meteorologia de la Armada (SOHMA), Uruguay . . 130
Servicio Geografico Militar, Uruguay 130
Servicio Geológico Nacional, Argentina 3
Servicio Hidrografico y Oceanografico de la Armada, Chile 23
Servicio Nacional de Aguas Subterraneas (SENAS), Costa Rica 32
Servicio Nacional de Geologia y Mineria "SERGEOMIN", Bolivia 12
Servicio Nacional de Geologia y Mineria (SERNAGEOMIN), Chile 23
Servicio Nacional de Meteorologia e Hidrologia, Bolivia 12
Servicios Hidrologicos y Climatologicos, Honduras 45
Servicos Geologicos de Portugal, Portugal 81
Shiga University, Japan 59
Shimane University, Japan 59
Shinshu University, Japan 59
Shirshov Institute of Oceanology (IOAN), Russian Federation 86
Shizuoka University, Japan 59
Sigma Gamma Epsilon, Epsilon Omega Chapter, United States 125
Sigma Gamma Epsilon, United States 125
Slovak Hydrometeorological Institute, Slovakia 89
Societe Nationale d'Operations Petroleres de la Cote d'Ivoire (PETROCI), Cote D'ivoire 33
Societe Nationale de Recherches et d'Exploitation Miniere (SONAREM), Mali 66
Societe pour la Realisation de Forages d'Exploitation en Cote d'Ivoire, Cote D'ivoire 33
Societe pour le Developpement Minier de la Cote d'Ivoire (SODEMI), Cote D'ivoire 33
Society for Archaeological Sciences, United States 125
Society for Geology Applied to Mineral Deposits, France 41
Society for Luminescence Microscopy and Spectroscopy, United States 125
Society for Mining, Metallurgy and Exploration Inc. (SME), United States 125
Society of Economic Geologists, United States 125
Society of Economic Paleontologists and Mineralogists, United States 125
Society of Engineering and Mineral Exploration Geophysicists, United States 125
Society of Environmental Geochemistry and Health, United States 125
Society of Exploration Geophysicists, United States 125

ORGANIZATION INDEX

Society of Explorationists in the Emirates, United Arab Emirates	103
Society of Geological Sciences, Romania	82
Society of Independent Professional Earth Scientists, United States	125
Society of Petroleum Engineers Inc., United States	126
Society of Professional Well Log Analysts, United States	126
Society of Resource Geology, Japan	59
Society of Vertebrate Paleontology, United States	126
Soil Science Society of America, United States	126
Sorby Natural History Society, United Kingdom	107
South African Institute of Mining and Metallurgy, South Africa	91
South Australian Department of Mines and Energy, Australia	8
South Carolina Geological Survey, United States	126
South Coast Geological Society Inc., United States	126
South Dakota Geological Survey, United States	126
South Pacific Applied Geoscience Commission, Fiji	38
South Texas Geological Society, United States	126
Southampton Oceanography Centre, United Kingdom	108
Southeast Asia Petroleum Exploration Society (SEAPEX), Singapore	89
Southeastern Geophysical Society, United States	126
Southern Cross University, Australia	8
Southwest Louisiana Geophysical Society, United States	126
Spanish Groundwater Club, Spain	93
Spanish Society of Paleontology, Spain	93
SPOT Image, France	41
St. Petersburg University, Russian Federation	86
State Commission for Dams, Iraq	51
State Committee on Geology, Uzbekistan	130
State Establishment for Geological Survey and Mining, Iraq	51
State Establishment for Surveying, Iraq	51
State Hydraulic Works (DSI), Turkey	102
State Oceanic Administration, China	30
State Seismological Bureau (IGSSG), China	30
State University "Lviv Politechnic", Ukraine	102
Stavropol Polytechnical University, Russian Federation	86
Student Geological Society, United States	126
Surtsey Research Society, Iceland	47
Survey and Land Information Department, New Zealand	73
Survey and Mapping Department, Somalia	89
Survey and Mapping Division, Tanzania, United Republic Of	98
Survey Department of Brunei, Brunei Darussalam	15
Survey Department, Jamaica	55
Survey Department, Myanmar	70
Survey Department, Nepal	71
Survey Department, Singapore	89
Survey Department, Zambia	133
Survey Division, Hong Kong	46
Survey of Bangladesh, Bangladesh	10
Survey of Egypt, Egypt	37
Survey of India, India	49
Survey of Kenya, Kenya	61
Survey of Pakistan, Pakistan	76
Surveyor General's Office, Sri Lanka	93

ORGANIZATION INDEX

Sustainable Land and Water Resources Management Committee, Australia	8
Swedish Environmental Protection Agency, Sweden	95
Swedish Meteorological and Hydrological Institute (SMHI), Sweden	95
Swiss Association of Petroleum Geologists and Engineers, Switzerland	96
Swiss Committee of the International Association of Hydrogeologists, Switzerland	96
Swiss Geological Society, Switzerland	96
Systematics Association, United Kingdom	108
Tahal-Water Planning for Israel and Tahal Consulting Engineers, Ltd., Israel	53
Taiwan Bureau of Mines, Taiwan, Republic Of China	97
Taiwan Provincial Water Conservancy, Taiwan, Republic Of China	98
Tartu University, Estonia	37
Technikon Pretoria, South Africa	91
Technische Universiteit Delft, Netherlands	72
Tektite Research, United States	126
Tel-Aviv University, Israel	53
Tennessee Division of Geology, United States	126
Texas A&M Geological Society, United States	126
Texas Bureau of Economic Geology, United States	126
Texas Water Resources Institute, United States	126
The Centre for Earth Science Studies, India	48
The Chartered Institution of Water and Environmental Management, United Kingdom	104
The Clay Minerals Society, United States	112
The Egyptian Geological Survey and Mining Authority, Egypt	36
The Geological Society, United Kingdom	104
The Geophysical Commission, Norway	75
The Kentucky Paleontological Society, Inc., United States	118
The Open University, United Kingdom	106
The Papua New Guinea University of Technology, Papua New Guinea	77
The Society for Organic Petrology, United States	125
The Surveyor General, Namibia	71
The Swedish National Road and Transport Research Institute, Sweden	95
The University Lancaster, United Kingdom	108
Third World Academy of Sciences, Italy	54
Tiefbauamt des Fürstentums Liechtenstein, Liechtenstein	63
TNO Institute of Applied Geoscience, Netherlands	72
Tobacco Root Geological Society, United States	127
Tohoku University, Japan	59
Tokai University, Japan	59
Tokushima University, Japan	59
Tokyo Geographical Society, Japan	59
Tokyo Institute of Technology, Japan	59
Tomsk State University, Russian Federation	86
Tongji University, China	30
Toronto Geological Discussion Group, Canada	21
Tottori University, Japan	59
Toyama University, Japan	59
Toyo University, Japan	59
Trinity College, Ireland	52
Tulsa Geological Society Inc., United States	127
Turkish Association of Petroleum Geologists, Turkey	102
Tver Polytechnical Institute, Russian Federation	86
Tyumen State Technical Oil and Gas University, Russian Federation	86

ORGANIZATION INDEX

U.S. Antarctic Marine Geology Research Facility, United States	127
U.S. Antarctic Research Program, United States	127
U.S. Bureau of the Census, United States	127
U.S. Committee on Irrigation and Drainage, United States	127
U.S. Department of Agriculture -- Forest Service, United States	127
U.S. Department of Commerce, United States	127
U.S. Department of Defense, United States	127
U.S. Department of Energy, United States	127
U.S. Department of Labor -- Bureau of Labor Statistics, United States	128
U.S. Department of Labor -- Mine Safety and Health Administration, United States	128
U.S. Department of the Interior -- Bureau of Indian Affairs, United States	127
U.S. Department of the Interior -- Bureau of Land Management, United States	127
U.S. Department of the Interior -- Bureau of Reclamation, United States	128
U.S. Department of the Interior, United States	127
U.S. Environmental Protection Agency -- Prevention, Pesticides, and Toxic Substance, United States	128
U.S. Environmental Protection Agency -- Solid Waste and Emergency Response, United States	128
U.S. Environmental Protection Agency -- Water, United States	128
U.S. Environmental Protection Agency, United States	128
U.S. Geodynamics Committee, United States	128
U.S. Geological Survey (USGS), United States	128
U.S. International Trade Commission, United States	128
U.S. National Committee for Geology, United States	128
U.S.Department of Labor, United States	128
Ufa State Oil Technical University, Russian Federation	86
United Nations Development Programme, Fiji	38
United Nations Environment Programme, Kenya	61
Universidad Autonoma de Guerrero, Mexico	67
Universidad Autonoma de Nuevo Léon, Mexico	67
Universidad Autonoma de Yucatan, Mexico	67
Universidad Autonoma de Zacatecas, Mexico	67
Universidad de Costa Rica, Costa Rica	32
Universidad de Guadalajara, Mexico	67
Universidad de Oviedo, Spain	92
Universidad Nacional Autonoma de Mexico (UNAM), Mexico	68
Universidade de São Paulo - USP, Brazil	14
Universidade Estadual de Campinas, Brazil	14
Universidade Estadual Paulista - UNESP, Brazil	14
Universiteit van Amsterdam, Netherlands	72
Universiti Sains Malaysia, Malaysia	65
University "St. Cyril and Methodius", Macedonia, Republic Of	64
University College Galway, Ireland	52
University College London, United Kingdom	108
University College, Cork, Ireland	52
University College, Dublin, Ireland	52
University of Aberdeen, United Kingdom	103
University of Adelaide, Australia	3
University of Arizona Geophysical Society, United States	110
University of Auckland, New Zealand	72
University of Birmingham, United Kingdom	103
University of Botswana, Botswana	13

ORGANIZATION INDEX

University of Bristol, United Kingdom 103
University of Cambridge, United Kingdom 104
University of Canberra, Australia 4
University of Canterbury, New Zealand 72
University of Cape Town, South Africa 90
University of Dar es Salaam, Tanzania, United Republic Of 98
University of Dhaka, Bangladesh 10
University of Durban-Westville, South Africa 90
University of Durham, United Kingdom 104
University of East Anglia, United Kingdom 104
University of Edinburgh, United Kingdom 104
University of Exeter, United Kingdom 104
University of Glasgow, United Kingdom 105
University of Helsinki, Finland 39
University of Hull, United Kingdom 105
University of Keele, United Kingdom 105
University of Leeds, United Kingdom 105
University of Leicester, United Kingdom 106
University of Liverpool, United Kingdom 106
University of London, Birbeck, United Kingdom 106
University of London, Imperial College, United Kingdom 106
University of London, Queen Mary, United Kingdom 106
University of Malaya, Malaysia 65
University of Manchester, United Kingdom 106
University of Mauritius, Mauritius 67
University of Melbourne, Australia 7
University of Mining and Geology, Bulgaria 15
University of Natal, Pietermaritzburg, South Africa 90
University of Natal, South Africa 90
University of New England, Australia 7
University of New South Wales, Australia 7
University of Newcastle Upon Tyne, United Kingdom 106
University of Newcastle, Australia 7
University of Otago, New Zealand 73
University of Paisley, United Kingdom 106
University of Papua New Guinea, Papua New Guinea 77
University of Petroleum, China 30
University of Port Elizabeth, South Africa 91
University of Pretoria, South Africa 91
University of Queensland, Australia 7
University of Rajshahi, Bangladesh 10
University of Reading, United Kingdom 107
University of Sheffield, United Kingdom 107
University of Sofia, Faculty of Geology and Geography, Bulgaria 16
University of South Australia, Australia 8
University of Southampton, United Kingdom 107
University of St. Andrews, United Kingdom 108
University of Stellenbach, South Africa 91
University of Stellenbosch, South Africa 91
University of Sydney, Australia 8
University of Tasmania, Australia 8
University of Technology, Sydney, Australia 8

ORGANIZATION INDEX

University of the Orange Free State, South Africa	90
University of the South Pacific, Fiji	38
University of the West Indies, Jamaica	55
University of the Western Cape, South Africa	91
University of Tokyo, Japan	59
University of Tsukuba, Japan	60
University of Waikato, New Zealand	73
University of Wales, Aberystwyth, United Kingdom	108
University of Wales, Bangor, United Kingdom	108
University of Wales, Cardiff, United Kingdom	108
University of Western Australia, Australia	8
University of Witwatersrand, South Africa	91
University of Wollongong, Australia	9
University of Zambia, Zambia	133
University of Zululand, South Africa	91
Urals State Academy of Mining Geology, Russian Federation	86
Ussher Society, United Kingdom	108
Utah Geological Association, United States	128
Utah Geological Survey, United States	128
Venezuelan Geological Society, Venezuela	131
Venezuelan Society for the History of Geological Sciences, Venezuela	131
Venezuelan Society of Petroleum Engineers, Venezuela	131
Venezuelan Speleological Society, Venezuela	131
Vermont Agency of Natural Resources, United States	128
Vermont Geological Society, United States	129
Vernadsky Institute of Geochemistry and Analytical Geochemistry (GEOKHI), Russian Federation	86
Vesuvian Vulcanological Observatory, Italy	54
Victoria University of Wellington, New Zealand	73
Vinogradov Institute of Geochemistry (IGKH), Russian Federation	86
Virginia Division of Mineral Resources, United States	129
Volcanologic and Seismologic Observatory of Costa Rica, Costa Rica	32
Volcanological Society of Japan, Japan	60
Voronezh State University, Russian Federation	86
Vyzkumny Ustav Geodeticky, Topograficky a Kartograficky, Czech Republic	34
Wadra Institute of Himalayan Geology, India	49
Waseda University, Japan	60
Washington Department of Natural Resources, United States	129
Water Affairs Department, Zambia	133
Water and Environment Research Office, Finland	39
Water and Power Development Authority (WAPDA), Pakistan	76
Water and Sewage Corporation, Bahamas	10
Water and Sewerage Section, Fiji	38
Water Data Unit, United Kingdom	108
Water Department, Kenya	61
Water Department, Libyan Arab Jamahiriya	63
Water Development Department, Cyprus	34
Water Environment Federation, United States	129
Water Research Center, Egypt	37
Water Resources and Dams Ministry, Libyan Arab Jamahiriya	63
Water Resources Authority, Jamaica	55
Water Resources Board, Sri Lanka	93

ORGANIZATION INDEX

Water Resources Branch, Swaziland 94
Water Resources Bureau, Japan 60
Water Resources Department, Iraq 51
Water Resources Division, Malawi 65
Water Resources Division, Sierra Leone 88
Water Resources Institute, Ghana 43
Water Resources Planning Commission, Taiwan, Republic Of China 98
Water Resources Research Centre (VITUKI), Hungary 47
Water Resources Section, Qatar 81
Water Supply Service, Suriname 94
Weather Forecast and Hydrology Service, El Salvador 37
West African Science Association, Ghana 43
West Texas Geological Society Inc., United States 129
West Virginia Coal Mining Institute Inc., United States 129
West Virginia Geological and Economic Survey, United States 129
Wisconsin Geological and Natural History Survey, United States 129
Women in Mining, United States 129
Working Group on Flood Basalt Volcanism, India 49
Works Department, Malta 66
World Coal Institute, United Kingdom 108
World Data Center A: Glaciology (Snow and Ice), United States 129
World Data Center A: Marine Geology and Geophysics, United States 129
World Data Center A: Oceanography, United States 129
World Data Center A: Solid Earth Geophysics, United States 129
World Glacier Monitoring Service, Switzerland 96
World Organization of Volcano Observatories, France 41
Wuhan Technical University of Surveying and Mapping, China 30
Wyoming Geological Association, United States 129
Wyoming State Geological Survey, United States 130
Yacimientos Carboniferos Fiscales, Argentina 3
Yacimientos Petroliferos Fiscales Bolivianos (YPFB), Bolivia 13
Yamagata University, Japan 60
Yamaguchi University, Japan 60
Yamanashi University, Japan 60
Yemen Geological Society, Yemen 132
Yemen Oil and Mineral Resources Corporation, Yemen 132
Yokohama National University, Japan 60
Yukon Territory Exploration and Geological Services Division, Canada 21
Zairian Geographic Institute, Zaire 133
Zairian Riverways Authority, Zaire 133
Zaklad Oceanologii PAN, Poland 80
Zambia Industrial and Mining Corporation, Limited (ZIMCO), Zambia 133
Zhingshan University, China 31
Zimbabwe School of Mines, Zimbabwe 134